JESUIT METEOROLOGISTS:

FORGOTTEN MEN

THE CASE OF MARC DECHEVRENS

First Director of the

Maison St Louis Observatory, Jersey

Frank Le Blancq

Société Jersiaise

PUBLISHED BY
Société Jersiaise
7 Pier Road
St Helier
Jersey
JE2 4XW
Channel Islands

www.societe-jersiaise.org

First edition

Jesuit Meteorologists - Forgotten Men: The Case of Marc Dechevrens
First Director of the Maison St Louis Observatory, Jersey
by Frank Le Blancq

Occasional Studies
Series number: 2

Text set in Sabon LT Pro
189x246mm
78 printed pages

Cover image: . From a portrait of Père Marc Dechevrens hanging in the
Maison St Louis Observatory. (Author's photograph)

Contents

Preface v
Introduction 1

Part One - The Jesuits 3
Part Two - Dechevrens: early life and career 7
Part Three - Dechevrens: in Jersey 25

Notes 47
References 57
Appendix - Works by Père Marc Dechevrens 61

Portrait of Père Marc Dechevrens which hangs in the Maison St Louis
Observatory. Each corner of the picture frame has a traditional character in
Chinese calligraphy known as seal script; in rough translation it reads 'Paying
great respect to Father Dechevrens'. Faint writing on the back of the wooden
frame appears to read on one strut 'Peinture faite a Tou se wè Chine. Depuis un
photograph'. On the other is 'Reçu en Novembre 1919'. (Painting made at Tou
se wè China. From a photograph. Received in November 1919). Tou se wè was an
orphanage at the Jesuit mission in Zi-ka-weï, where the youngsters were taught
trades such as carpentry, painting and printing. (Author's photograph)

Preface

The Jesuits made important contributions to the development of science, including meteorology, particularly by establishing observatories in various countries during the 19th century including Jersey. However, over the years the contribution of the observatories and the highly intelligent and dedicated scientists who worked in them have been neglected and largely forgotten. This work aims to put the principles of the Jesuit order into scientific context, then outlines the career and contribution to meteorology of one individual in particular, Père Marc Dechevrens. He arrived at the Zi-ka-weï Observatory in Shanghai, China in 1873 and was its Director from 1875 until 1887.[1] Later he came to Jersey where he was Director of the Maison St. Louis Observatory for the 30 years from 1894 until 1923.

This work is a revised and much expanded version of an unpublished paper written in 2005. The old photographs have been digitised and processed to show them to best advantage by cropping, adjusting exposure and retouching to remove staining and blemishes where possible.

Frank Le Blancq
29 November 2021

Introduction

At a meeting convened in early 1993, staff at the Jersey Meteorological Department (JMD) agreed to commemorate the centenary of the Observatory at Maison St Louis, Jersey where the first weather recordings were made by its Director Père Marc Dechevrens on 1 January 1894.[2] A small group of JMD staff undertook to organise a celebration on 23 March 1994, the day chosen to coincide with the annual World Meteorology Day. A plaque was unveiled by Jersey's Bailiff, Sir Peter Crill, in front of representatives from the UK Met Office, Météo-France, the Jesuits, the resident observer and invited guests.

In the course of preparations for the centenary a good deal of original material dating back to the early years of the Observatory was discovered in cupboards, draws and boxes. The author of this work was aware that some of it was of historical interest and importance, and made copies of papers and documents relevant to the history of the building and its weather recorders in the previous 100 years. The information was used to write a brief history of the Observatory (Le Blancq 1994) and later a more detailed study of the life and work of Marc Dechevrens (Le Blancq 2005). That information is greatly expanded in this work, because much additional material has come to light in subsequent years, particularly through the medium of the internet, where new items appear regularly and are worthy of placing on record.

The text is split into three parts. First, the author examines the Jesuits in the world of science and scientific development, to place their ideas and working methods in perspective. The background to the Jesuit order and the situation in China are then explained. In the second part he concentrates on the life of Marc Dechevrens in China to put his work in context, for he was more than a simple weather recorder, important though that was. He was one of many Jesuits who dedicated their life to science and religion and became well known in the Far East, where he set up a typhoon warning service along China's east coast. We shall see that he made meticulous weather recordings in Shanghai, researched tropical storms and debated their likely structure with leading meteorologists of the day. The third part concerns the later phase of his life in Jersey, where he founded and for nearly 30 years operated the Maison St Louis Observatory, as well as remaining actively engaged in various lines of research. Finally a list of Dechevrens' known works is presented as an appendix.

Extensive notes are included at the end of the main body of this work and are intended to be read in conjunction with the text. Some may appear somewhat incidental to the main subject, but are included to provide greater detail about Jesuit scientists in general, their working methods, their deep involvement in the development of science and the strong connections that existed between Jesuit meteorologists dispersed in the far corners of the world.

Part One

The Jesuits and their principles

I do not desire any other reward, besides the one I expect from
God, than being useful to my brethren and to contribute in some
way to the advancement of science and the welfare of humanity.
~ Benito Viñes, Jesuit meteorologist

The Society of Jesus - the Jesuits - have always held learning in high regard, having principles which strongly encourage active engagement with the world, that is to say the world outside religion. The foundation of this Catholic order by Ignatius of Loyola in 1534 (and sanctioned Pope Paul III in 1540) came during the Renaissance. Expansion and promotion of education by the Jesuits coincided with the birth of modern science and they played an active role in its development. However, we need to recognise that while Jesuit scientists went about their studies, mainly astronomy, natural sciences and mathematics in the early days, they lived in an intense intellectual community amongst their philosopher and theological colleagues; life within the order was not always without tension and conflict. To illustrate this point we may cite the Galileo affair as an important example. In 1611 Jesuit scientists at the Collegio Romano embraced Galileo's ground-breaking observations which concluded that the earth orbited the sun (heliocentrism), as proposed by Copernicus nearly 100 years earlier. This acceptance came despite a clear conflict with literal interpretations of the Scriptures, which fix the earth at the centre of the universe, around which the universe revolves (geocentrism). However, just five years later in 1616, Galileo was forbidden by the Church from advancing the heliocentric theory and effectively silenced. Following the publication of a book, he was tried again in 1633 and held under house arrest for the rest of his life. Thus, Jesuit scientists having accepted heliocentrism, soon found their decision reversed. In an intellectual clash between strong Jesuit personalities, the scientists recognised the threat their views posed to religious doctrine and backed down, unwilling to face the implications of supporting a true scientific revolution. The arguments over this issue are discussed in some detail by Coyne (2015) and Udías (2003).

Despite the Galileo affair, it is arguably reasonable to say that Jesuit scientists were in general less hindered by the medieval traditions which afflicted the Catholic Church at the time. Following restoration of the Jesuits after their suppression from 1773 to 1814, highly trained members were scattered to all corners of the world. Striving 'to find God in all things' they were selfless, self-disciplined, dedicated to duty and sought no personal reward as well stated in the Viñes quote above. The order accepted science as a means of fulfilling their aims through education and missionary work. In this way they contributed to the advancement of science, but also the well being of communities in a world where science was often seen as hostile to religion; as time went on it showed the two were compatible. The order made important contributions to the development of several branches of science. From the beginning astronomy, mathematics and the natural sciences were fields of study in the colleges and universities they established in Europe and later overseas. As new sciences emerged, such as seismology, so Jesuit scientists studied them and usually to a high intellectual level. Some became university professors and a few remain so in the present day. The reader is referred to Udiás (2003) for details and further discussion on this subject.

Jesuit observatories and meteorologists

As mentioned above, an important aspect of Jesuit activity was establishing observatories, particularly in the 19th century. Some became important centres of research and learning, especially in the fields of astronomy, geophysics and meteorology. However, to a large extent the observatories and the men who served in them have become a rather forgotten and certainly a neglected chapter in the history of those sciences. In more recent years Udiás (1996 and 2003) has attempted to redress the balance by discussing the role of Jesuit meteorological observatories. Udiás and Stauder (1996) have also outlined their contribution to seismology, while Ramos Guadalupe (2014) has related the interesting life of Benito Viñes, a Jesuit pioneer in tropical meteorology.

From the mid-19th century, twenty-nine observatories devoted mainly or exclusively to meteorology were established in twenty countries. In some of those countries in Africa, South America and Asia they were the first of their kind. A complete list can be found in Udiás (2003), but those at Belén (Cuba), Manila (Philippines), Tananarive (Madagascar), Ksara (Lebanon) and Zi-ka-weï (China) were probably the most important. It is also generally overlooked that Jesuits developed the national weather services in Colombia, the Philippines and Madagascar.

Many Jesuit priests working in the observatories had long and distinguished careers often spanning several decades. This work looks at Père Marc Dechevrens' 61 years in the Jesuit order, examining in particular a period of over 40 years when he was involved in the development of two observatories and in more theoretical aspects of meteorology as a science.

Part Two

Dechevrens' early life and his move to China

Marc-Antoine Dechevrens was born on 26 July 1845, at Chêne-Bourg near Geneva in Switzerland and raised with an older brother and a sister, both of whom joined religious orders. Influenced by his older brother, in 1862 he joined him in the Jesuit order as a novice. French was his first language and throughout his life he was associated with the French province of Jesuits, though we know through correspondence in the Shanghai and Jersey newspapers that he was competent in the English language.

In an appreciation of his life, Gautier (1924) tells us that from 1862 the young Dechevrens studied at the Fribourg Seminary in Switzerland. In the Jesuit teaching tradition he then taught physics from 1869 to 1872 in Jesuit colleges at Vannes in Brittany and Vaugirard in Paris. While in that city he attended a course in mathematics at the Sorbonne. Despite his young age, it was not long before superiors in the order recognised his potential. In mid 1873 he was approached and quickly accepted the offer of a post in the meteorological observatory at Zi-ka-weï, near Shanghai, which had been established just two years earlier. However, before the long sea voyage to China, Dechevrens was sent to Stonyhurst College in Lancashire, which was used as the Jesuit training ground for future heads of observatories. For over two months he received training in meteorology and observatory work, under the guidance of Stephen Perry, one of the foremost Jesuit scientists of the day.[3]

Setting the weather scene

Prior to the 19th century, the knowledge the captain of a ship acquired about weather and ocean conditions such as winds, waves and currents was not for general broadcast. O'Brien (2014) neatly sums it up: 'To best their competitors they jealously guarded the data they amassed - it was secret'. This secret knowledge represented a commercial advantage for ship owners, be it in the China Seas or elsewhere. However, as the 1800s progressed and trade expanded, the need for change became increasing apparent, not least in the face of regular destruction caused by both mid-latitude and tropical storms.[4] In the China Seas, the main fear was typhoons; they caused death on land and at sea; they wrecked and sank ships; they incurred losses for insurance companies underwriting the ships and their cargoes; they jeopardised the livelihood of shipowners, merchants and fishermen; port authorities suffered economically. In short, typhoons in the China Seas were the serious common enemy of many and came to play an important role in advancing weather research in general, through the need for a storm warning system. Unfortunately, decades of debate by meteorologists over conflicting theories of storm genesis and storm structure through much of the 19th century, held back these advances.

The first steps towards storm forecasting services were made in Europe in the mid-19th century. As a start, the Beaufort wind scale, Luke Howard's cloud classification and Henry Piddington's standard storm vocabulary had already been widely adopted by the British (Williamson and Wilkinson 2017) and were being adopted by other nations.[5] Then in 1853 the First International Maritime Conference convened in Brussels and initiated a standard method for reporting weather at sea in ship logbooks using a common set of instructions. At the Conference a strong case was also made for the use of properly certified instruments with the aim of improving accuracy.[6]

It was two maritime disasters in particular that pushed France and Britain into setting up systems to warn of bad weather along European coasts and over sea areas. They were the Black Sea storm in November 1854, when 37 ships were wrecked or seriously damaged and the 'Royal Charter' storm on the Welsh coast at Anglesey in October 1859, when about 450 people drowned and many ships wrecked. On land at least, merely recognising the signs of an approaching storm was not enough, a network of observers was needed with a means of exchanging weather information and warnings. In this respect technical innovation in the form of the electric telegraph played a vital role. The new telegraph networks over land provided a quick means of communicating information from the 1850s onward. A network of underwater cable connections soon followed, though it should be noted that both were subject to frequent disruption for various reasons including cost, poor maintenance, politics and even bad weather!

The situation in the China Seas

The starting point in China came some years later. It was in 1869 that Robert Hart, Inspector-General of the Chinese Imperial Maritime Customs, initiated a project with the provision of storm warnings in mind (Zhu 2012). Hart's plan foresaw ports and lighthouses along the coast fitted with weather-recording instruments and connected to the telegraph network, with Peking acting as a central collecting observatory to receive their reports. Progress was very slow and in an attempt to advance his plan, in 1873 Hart wrote to many authorities extending geographically from Vladivostok to Singapore. Unfortunately his relationship with China's mercantile community was poor and without someone to co-ordinate and implement his project it remained a plan. In 1875 Labrosse wrote that '. . . a typhoon is always to be feared . . . hardly any precaution can be considered absolutely efficacious for avoiding their effects' (Labrosse 1875: 29). This was the state of play when Marc Dechevrens became Director of the Zi-ka-weï Observatory in 1875. The need for a warning service was very evident, but the China coast had to wait almost another decade before it came about.

Dechevrens arrives in China

Shanghai was a treaty port, that is a port open to foreign trade and residence, when Dechevrens arrived in the French concession (itself established in 1849) and Zi-ka-weï was a village about 10 kilometres south-west of the port area. There in a 1½ square mile enclave the Jesuits

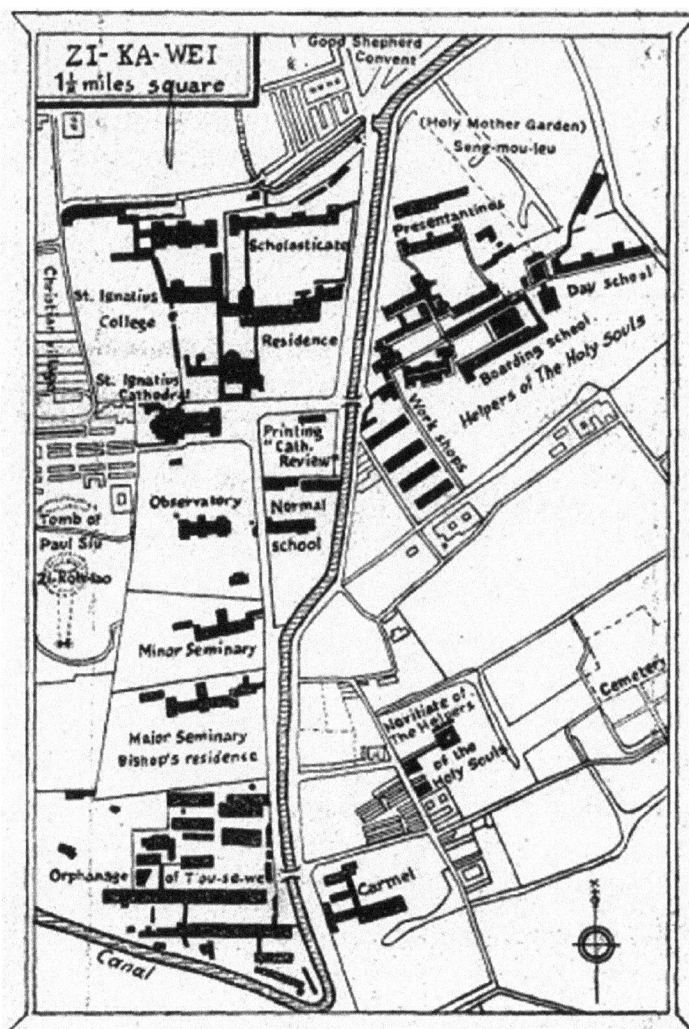

Figure 1: A plan of the Jesuit enclave at Zi-ka-weï of unknown date. The Observatory is at centre left, to the south of the Cathedral. (Source unknown.)

ran a boarding school, seminary, printing press and two orphanages, but now it lies well within the modern city of Shanghai.

Dehergne (1976) noted that Henri Le Lec was the first Jesuit to take weather observations in Shanghai in 1865, but it was not until December 1872 that a continuous set of recordings began at the new observatory.

Figure 2: A view of Zi-ka-weï Observatory from the south, believed to be taken in the early 1880s. Note the second storey and a prominent 33-metre wooden tower built to house Dechevrens' anemometers well clear of the ground. (St Louis Observatory archive)

Figure 3: Zi-ka-weï Observatory viewed from the north, believed to be taken in the early 1880s. Magnetic recordings were made in the octagonal building at lower right, which was built without metal parts to avoid electrical interference. (St Louis Observatory archive)

When Dechevrens arrived at Zi-ka-weï a year later he joined the observatory as an assistant, attending to magnetic recordings as well as meteorological readings.[7] Soon after, in 1875, he was appointed Director and from that time under his leadership, developed the observatory's true scientific character as an institute of high standards and quality research focused on typhoons. He soon had a complete set of weather instruments in operation including a sophisticated Secchi meteorograph. This was an early form of automatic weather station, simultaneously recording pressure, temperature, humidity, rainfall, wind speed and wind direction, which would have eased the weather recording routine.[8] To the existing observatory building Dechevrens had a second storey added and on land to the south a 33-metre wooden framed tower was built to house anemometers. For this he designed a special 'clino-anemometer' and started a long study on vertical air currents which he later continued in Jersey.

Figure 4: A 180 lire stamp issued by the Vatican City in 1979 to commemorate the death of Angelo Secchi in 1878, includes a diagram of his meteorograph, an early form of automatic weather station. (Credit: Smithsonian National Postal Museum)

12

The 1879 Typhoon report

Marc Dechevrens studied typhoons from both a practical and theoretical point of view and here we concentrate on the practical.[9] As a meteorologist in China, he would have been well aware of their destructive power which periodically wreaked havoc along the coast and over the China Seas. Indeed, from time to time he provided information about them to the Shanghai newspapers. He would also have been aware of the research being carried out by fellow Jesuit Benito Viñes on tropical storms in Cuba, of his success at forecasting storms over the island and possibly of Viñes' vision of a Caribbean storm warning network.[10] Dechevrens' interest was probably heightened when a typhoon caused damage around Shanghai on 31 July 1879. He undertook a detailed study of this storm by gathering a vast number of observations and reports from land stations and ship logbooks, which he used to compile an exhaustive report. It

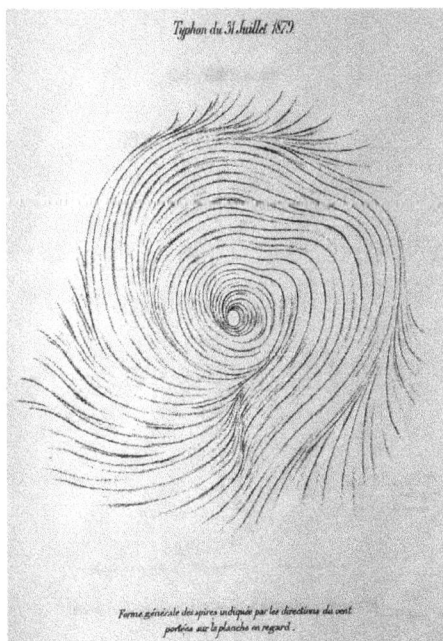

Figure 5: A page from Dechevrens' 1879 typhoon report showing a sketch of his idea of the form of a typhoon. (St Louis Observatory archive)

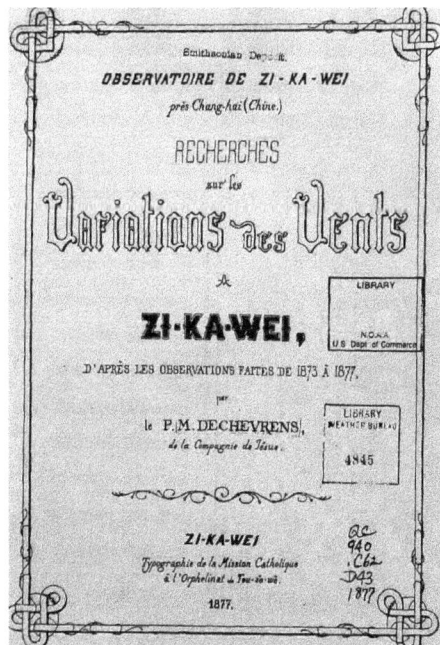

Figure 6: One of Dechevrens' earliest papers concerning his research on the variation of winds at Zi-ka-weï. Source: Author's image.

included an analysis of the typhoon's track and a draft of what was to become his hydro-dynamic theory on cyclone formation. His report - *Le typhon du 31 Juillet 1879* - was published (Dechevrens 1879) with an abbreviated version in English, which drew the work of his Observatory to the attention of the local authorities and the international community in Shanghai.

From around this time, the study and forecasting of typhoons slowly evolved as the Observatory's main task, which continued long after Dechevrens departed. He had already been at work developing a network of stations that sent weather reports to Zi-ka-weï including, on an informal basis, some from Chinese Customs ports. His network depended on collaboration with Chinese and foreign telegraph companies, who agreed to waive their fees, and with mail services and shipping companies. He also received regular reports from the captains of over forty sail and steam ships trading in the China Seas. A map in the 1880 Observatory's annual bulletin shows a total of 54 stations in contact with Zi-ka-weï, though their reports were by no means all received in good time, or necessarily daily. Significantly, on the map Zi-ka-weï is shown as the 'Central Observatory' which Zhu (2012) suggests points to Dechevrens' ambitions for his weather centre. It was certainly becoming the *de facto* central observatory in China, which Hart had hoped to achieve for Peking.

Around 1880, local newspaper editorials, letters and comments, which advocated the need for storm warnings to mitigate against damage and losses, were in keeping with Dechevrens' ambitions. That year he wrote a 13-page memoir for observers entitled 'Meteorological service on the coast of China. Instructions for keeping the meteorological log'. Then in the winter of 1882 he embarked on a trip to France and returned with seventy sets of certified instruments to equip land stations and ships; as a side issue this led to controversy over metrification. Clearly our priest-meteorologist in Shanghai was very active with plans for the Observatory.

Mercantile interests

An interesting outsider's view of the development of the Zi-ka-weï weather service at this time is given by Zhu (2012) who, in a detailed thesis, studies the situation from the point of view of port users along the coast. He emphasises the importance of an 'inter-port group' (IPG), being a collection of like-minded influential men with a common interest in commerce and trade. This small group, comprising largely of expatriates, circulated and discussed information, especially weather information, through local newspapers such as Shanghai's *North China Daily News*, some of which they controlled. The press was their easiest channel of communication and by way of letters and editorials, essentially acted as the mouthpiece of the mercantile community. In the eyes of the IPG, the *raison d'être* of a weather service must be to forecast typhoons, as that is what served their interests and they highlighted the rise of Dechevrens and his private observatory (Dehergne 1976). It is perhaps also worth noting that the IPG and the Jesuits were both expatriate groups, providing a common bond of sorts.

Following newspaper pressure, in September 1881 the Shanghai Chamber of Commerce convened a meeting to discuss the feasibility of organising a system of meteorological reports with the aim of issuing warnings. Those present supported the proposal and Dechevrens was invited to co-operate, with mention of subsidies and subscriptions as part funding. He responded to the Chamber with a ten-point proposal to set up the service; our energetic Jesuit meteorologist was ready to proceed with the China Coast Meteorological Service (CCMS) much as originally foreseen by Robert Hart.

In fact at this time Hart, head of the Chinese Customs, was a sticking point. He was not on good terms with the mercantile community, but his co-operation was needed for the service to succeed, though in essence it replaced his own plan. However, more pressure in the local press led to his formal agreement in May 1882, whereby Chinese Maritime Customs stations would exchange weather reports with Zi-ka-weï and thus the CCMS scheme moved forward. From the weather reports received, Dechevrens was able to plot charts to detect and forecast weather features including the movement of typhoons and by 1884 reporting stations extended from Siberia and Japan, through north, central and southern China, to the the Philippines and 'English Indies', probably referring to Singapore (Zurcher 1884). From 1882 the Observatory also supplied daily weather information and a 24-hour forecast to several Shanghai newspapers.

These initiatives and the collaboration required, all point to Dechevrens' competence and organising ability. Organising the service will have entailed a lot of correspondence and much administration, as well as a good degree of tact and diplomacy. He did encounter problems a little later when, for instance, the querulous Dr William Doberck, Director of the Hong Kong Observatory, attempted to sidestep Zi-ka-weï's work and interfered in its relations with the Jesuit observatory in Manila.[11]

Despite the time-consuming regular routine of weather observation, co-ordination of the Observatory's annual summary and his research (outlined below), in 1884 at the invitation of the French municipality Dechevrens initiated two public services in Shanghai, which became of great importance to China Sea mariners. First, a semaphore mast was erected at the port on the French Quay where visual typhoon warning signals could be hoisted in sight of mariners and port users. Dechevrens wrote that he designed ten warning symbols to indicate the presence and direction of typhoons (later improved by a successor, Louis Froc) which were adopted by the Chinese Maritime Customs stations. The semaphore displayed other weather information too, such as wind conditions at the Gutzlaff telegraph station at the mouth of the Yangtze River, about 100 kilometres from the port of Shanghai (Tissandier 1891).

Second, a time service was established for adjusting chronometers, those vital timekeepers for determining the accurate position of a ship at sea. Until then a midday gun had been fired from the English concession, but the service was deemed unsatisfactory. The Chinese government funded a telegraph line from Zi-ka-weï to the port, while the municipality built a second semaphore mast from which a wicker sphere, the time ball, was released. Observations of the sun were made using a meridional telescope at the Observatory to determine time precisely and at exactly

Figure 7: The time ball on the quay at Shanghai port would have looked similar to that seen here at Greenwich Observatory in 2015. The ball was raised to the top of the post and dropped at exactly midday. (Author's photograph)

midday the time ball was released by an electric signal from Zi-ka-weï. Dechevrens inaugurated the service on 1 September 1884 (Dechevrens 1920). These developments again, amply demonstrate the Director's ability at organising and co-ordinating with various institutions.

Typhoon research

Apart from setting up and operating a typhoon warning service, Dechevrens was also busy researching a wide range of subjects. He made a considerable contribution to meteorological theory; indeed it started in earnest with his theory of cyclone formation in the 1879 typhoon report. He supported it using weather observations and data gathered from experiments he carried out using a barrel adapted to show the rotational nature of fluid flow (Dechevrens 1879 p.7). Essentially his theory recognised a lower part of the atmosphere, where air converged towards the centre of a cyclone and an upper region, from which air descended into the cyclone. Influential scientists at the French *Académie des Sciences*, including Hervé Faye, quickly dismissed his theory. Dechevrens was clearly annoyed and a little later accused the Academy of having 'preconceived ideas' (*taxée d'idées préconçue*) (Dechevrens 1881). As supporters of his ideas Dechevrens cited Erwin Knipping, a German meteorologist working for the Japanese weather service, who had deduced a similar theory, and other well-known researchers including fellow Jesuit Benito Viñes, who was studying hurricanes at Belén in Cuba. Dechevrens' annoyance was not without foundation.

A few years later, in a review of storm knowledge, Rosser (1886) noted that conjecture concerning the character of tropical storms had been replaced by a more systematic enquiry, as was being practised by Viñes, Dechevrens and Knipping who were building on the knowledge of earlier researchers. However, Rosser was scathing in his criticism of Faye at the Academy, who clung to his old theories. Rosser regarded Faye's defence of his proofs as 'rather a good joke' and commented more diplomatically that 'the views of French meteorologists and navigators do not coincide with those of M[onsieur] Faye' (p.101).

Cyclone theory in the 1870s and 1880s

The development of cyclone theory in the 19th century has been covered comprehensively by Kutzbach (1979). In the 1870s and 1880s a thermal theory of cyclones was widely accepted, whereby the release of latent heat during condensation in ascending currents of warm, moist air was considered the main source of energy for the formation and maintenance of cyclones. As a result, cyclonic disturbances in the atmosphere were believed to have a warm core. However, a fundamental problem was the lack of vertical temperature data with which to verify the theory. At this time we should note that the structure of both mid-latitude and tropical cyclones were thought to be similar, except that those in the tropics produced stronger winds.

On a practical level the exchange of ideas and discussion between scientists, was rather slow compared with the modern era which sometimes resulted in researchers developing similar ideas independently. Indeed, in the 1870s specialist meteorological journals were in their infancy. Dechevrens was somewhat isolated working in China where he developed his cyclone theory but, as his 1879 typhoon report indicates, he was certainly familiar with the ideas of other researchers and the literature of the time, citing papers and books written by the leading figures.

A digression - the work of Henri Peslin

An article by Rochas (2005), reminded us of French mining engineer Henri Peslin's important contributions to meteorological theory. Peslin published papers in 1868 and 1872, recognising the existence of ascending vertical currents in cyclones and developed the idea of stability and instability, a fundamental concept in understanding meteorology. However, his ideas were not well received by the leading scientists at the *Académie des Sciences*, Urbain Le Verrier and Hervé Faye. In 1875 Peslin used the results of his earlier works to counter a tenuous theory advanced by Faye, which proposed that vertical currents in cyclones were descending and not ascending. Rain could not be explained without ascending currents, Peslin argued, which allow condensation to occur. In the debate, conducted on the pages of the Academy's *Comptes Rendus*, other scientists, particularly Hildebrandsson, supported Peslin but he was ostracised by the powerful heavyweights at the Academy.[12] He made no further contribution to meteorology and disappeared from the scene, remaining little-known in meteorological circles.

Clearly, despite discovering an important meteorological concept, powerful players could easily smother ideas which did not conform to their own. It is very likely that Dechevrens knew of Peslin's treatment, so he was probably not surprised that his own theory advanced in 1879 received the same dismissive treatment, prompting his comment of 'preconceived ideas'. Fortunately, unlike Peslin, the Academy's negative response did not deter Dechevrens from publishing further research and later he made specific reference to Faye's theory (Dechevrens 1888).

Dechevrens' cyclone theory

To test his cyclone theory, Dechevrens studied the vertical pressure and temperature structure of the atmosphere, using published data recorded simultaneously at mountain summit stations and nearby low level stations in France and the United States. In two important papers (Dechevrens 1886 and 1887) he showed that when pressure was low at the base and summit stations, temperatures were significantly above average at the base but below average at the summit. In a comment at the end of a translated version of his 1886 paper in the American Meteorological Journal, the editor wrote 'We entirely agree with M. Dechevrens' result for pressure "in the height" though it should be noted that **this is directly contrary to theory** . . .' (author's emphasis).

Concurring nearly a century later, Kutzbach (1979) in her thorough review of 19th-century cyclone theory, remarked that Dechevrens' findings, aside from the theory he associated with them, were 'quite sensational because they conflicted with the thermal theory' (p.137). From that time the validity of the then accepted thermal theory of cyclones was challenged. Leading Austrian meteorologist, Julius von Hann, had made observations of the vertical temperature structure in anticyclones (rather than cyclones) in 1876.[13] Though they received little attention at the time he soon confirmed Dechevrens' findings with further data from Austria. In a paper written in 1893, but published in English much later (Hann, 1914), he showed that within five years of Dechevrens' findings, the convective or thermal theory was unsustainable. It seems therefore, that Dechevrens' discovery played an important part in changing and advancing the course of cyclone theory.

Air currents in the upper atmosphere

Dechevrens' broad-ranging research at Zi-ka-weï also included investigation of air currents in the upper atmosphere, with data recorded from at least as early as January 1877. He did this by observing the direction of movement of cirrus cloud found at altitudes of 20,000 to 30,000 feet. A summary table of the results included in the 1879 Shanghai typhoon report (p.5) revealed a dominant west-to-east movement for most of the year. At that time Englishman William Clement Ley, a master in the field of cloud studies was also making detailed observations of cirrus clouds in a slightly different connection (Kington 1999) but there is no mention that Dechevrens was aware of Ley's work. Several years later he wrote a more detailed paper (Dechevrens 1885) on upper air currents determined from the movement of cirrus, later cited by Hildebrandsson and Teisserenc de Bort (1903) as the first demonstration of the mainly west-to-east movement of upper winds in temperate latitudes.[14] In the same paper Dechevrens also advanced a theory linking westerly airflow in the upper atmosphere with north-east trade winds in the lower atmosphere, which he believed to be the same currents descending over the ocean.

Dechevrens' achievements in China

Whatever the merits of his cyclone theory in the light of current knowledge, Marc Dechevrens was clearly at the cutting edge of meteorological research at the time. His work on the vertical temperature structure of cyclones and movement of upper air currents can be said to be pioneering, leading to the advancement of meteorological theory by others. However, as we have seen, he was not only a theoretician but also a very practical meteorologist, developing the first private weather service. The typhoon warning service, port signals and time service were of great benefit to sailors, merchants and the general community who were very appreciative of the services emanating from the Observatory. Dechevrens' name and those of his successors were held in high regard in the Far East. In 1887 he was elected a member of the prestigious Jesuit scientific body, the Accademia Pontificia dei Nuovi Lincei, but his stay at Zi-ka-weï came to an abrupt end that year.

During his time in Shanghai, Dechevrens wrote many other meteorological papers, reports and memoranda on a wide range of topics and his early writing would surely have given him good standing on the

Figure 8: A later view of the Zi-ka-weï Observatory rebuilt in 1906, probably taken in the 1920s. (Image courtesy of Gaston Demarée, Royal Belgian Met Service)

Figure 9: The Zi-ka-weï Observatory building as seen in 1995 when it was used as offices, but trees largely obscured the view. (Author's photograph)

developing meteorological scene. He started in September 1874 with the Zi-ka-weï Observatory monthly meteorological and magnetic reports and an annual summary. Margollé (1879 and 1880) gave an indication of the thoroughness and range of his reports. For instance, that of 1877 discussed a heavy fall of sand and dust over the area, ocean undulations following an earthquake in Peru, crepuscular rays and Formosan typhoons. The 1878 report commented on meteorological factors relevant to the great famine in China, when an estimated 10 million people died. Other topics covered included zodiacal light (1875), magnetic variation during eclipses (1883), and the climate of Shanghai (1885). In addition he published a comprehensive annual report of typhoons in the China Seas (in 1881 it ran to 171 pages) investigating eighty individual storms over his fourteen years in China (Dechevrens 1920).

Dechevrens leaves China

Doubts over the circumstances of Dechevrens' departure from Zi-ka-weï have been raised by Zhu's (2012) evaluation. He postulates that Dechevrens left Shanghai due to several factors which became too heavy a burden for him. They include criticism he encountered over problems with the time ball service. In reality they were probably due to a poor telegraph connection, as the final few kilometres of the line from the port ran through a Chinese part of the city where the line was sometimes vandalised. It was known to be a weak link. There was controversy in the local press over the failure of Zi-ka-weï to predict a typhoon in August 1886; as in modern times the press could be a powerful enemy as well as a friend. Additionally, some of the Zi-ka-weï's reports may have been distorted or misrepresented by the querulous Dr Doberck, Director of the Hong Kong Observatory, who regarded the Jesuits at Zi-ka-weï (and Manila) as rivals rather than collaborators. Further, there was controversy over metrication, as the instruments Dechevrens brought back from France in 1883 were marked in metric units. One might regard this latter as a minor issue, but until that time Imperial units had been the standard and predictably it provoked criticism from the British contingent in

Shanghai. All new projects encounter problems in their early phases and the CCMS was no exception. Only as a footnote does Zhu add 'it was said at a much later time that he was sick in 1887 which may be why he left Shanghai' (Zhu 2012, p.111).

While the factors cited by Zhu may have had a bearing on the matter, Dechevrens was very clear that he suffered a severe attack of dysentery in September 1887 requiring a period of rest and recuperation, which was the reason for his return to Europe in October (Dechevrens 1920). It is not in the Jesuits' makeup to submit to outside pressures and as Jesuits take a vow of poverty, chastity and obedience on joining the order, he would have 'obeyed' a decision to leave Shanghai made by his superiors. Following his return to Europe, the Zi-ka-weï Observatory continued forecasting typhoons for decades under eminent successors, so one must at least raise a query over Zhu's interpretation of his departure. Regardless of whether there was more to his departure than health, Dechevrens had reason to be proud that under his guidance the Observatory was established on a sound footing, an institute of high standards and high quality meteorological research.

Part Three

A new life in Jersey

As brief background information, in 1880 the Jesuits set up a training college at the Maison St Louis in Jersey, Channel Islands. This came about because they had effectively been expelled from France at very short notice when decrees of the education minister Jules Ferry were quickly enacted in the French parliament (Le Blancq 2005). The laws secularised education, forbidding religious orders from directing or teaching in any educational establishment, so to continue their teaching activities the Jesuits had to leave France in a great rush. They purchased the former Imperial Hotel in St Saviour's Road, with land extending back over the high ground to the top of Highlands Lane and named the property Maison St Louis.

Dechevrens' whereabouts on his return to Europe are not entirely clear. According to Stierli (2005) he was in Jersey teaching natural sciences at Maison St Louis from 1888 to 1891, though he was not in the Island for the Census on 6 April 1891. Then for a short time from 1891 to 1893 he was teaching in Constantinople. He returned to Jersey in October 1893,[15] noting that he was 'restored to full health' (Dechevrens 1920), thereby implying that it took a while to completely recover from the attack of dysentery which sent him back from China. That year, according to Delattre (1953, p.859), Dechevrens sought and was granted permission to build an observatory at Maison St Louis with necessary funds allocated by his superiors, suggesting his return to Jersey was planned. His three main aims on becoming Director of the observatory were to take general weather observations, undertake research and train young Jesuits for work in the Jesuit observatories overseas. It was a thoroughly French establishment in personnel, language, instruments and observing methods.

The Maison St Louis Observatory and its work

In the first edition of the Observatory's annual bulletin for the year 1894, Dechevrens notes in the introduction that it was a busy year. He also expressed his hope that he would find a welcome and trust from other observatories and learned societies as he had at Zi-ka-weï. Having set up instruments at the end of 1893, his first weather observation was taken on 1 January 1894. This was just before builder Samuel Cuzner began construction of the Observatory building on high ground near the top of Highlands Lane, some 55 metres above sea level. Designed and built under Dechevrens' guidance, it was completed in October 1894 at a cost of £531, comprising a hallway and eight rooms, including a transect room on the east side to establish accurate time using sun sightings. As work on the Observatory progressed, a 50metre (164 feet) steel tower was being erected alongside it and completed in November.[16] The tower was for Dechevrens to continue studying horizontal and vertical wind currents that he had begun in Zi-ka-weï, well clear of turbulence caused by surface obstructions. It was a highly visible structure, forming a very impressive landmark on the hill with a profile akin to the Eiffel Tower and known locally as 'the Jesuits' tower'. In China, Dechevrens had built a special 'clino-anemometer' to record wind velocity and had a more complex instrument of his own design constructed for the new tower by *Richard Frères* in Paris. Dechevrens' *grand anémomètre-universal* was placed atop the tower in February 1895, some 112 metres (367 feet) above mean sea level. Apart from his anemometer in Jersey, for a time another was working on top of the Eiffel Tower and one was known to be in Washington for a short time. Dechevrens spoke about it at the *Exposition Universelle de Paris* in 1889 and the Jersey instrument gave good service, but in practice the mechanism was too complicated for widespread adoption.[17]

On completion, Maison St Louis Observatory was equipped with a range of standard weather recording instruments and others were added from time to time, which Dechevrens noted in the annual reports. In 1895 a Piche evaporometer and a very sensitive air and water micro-barograph

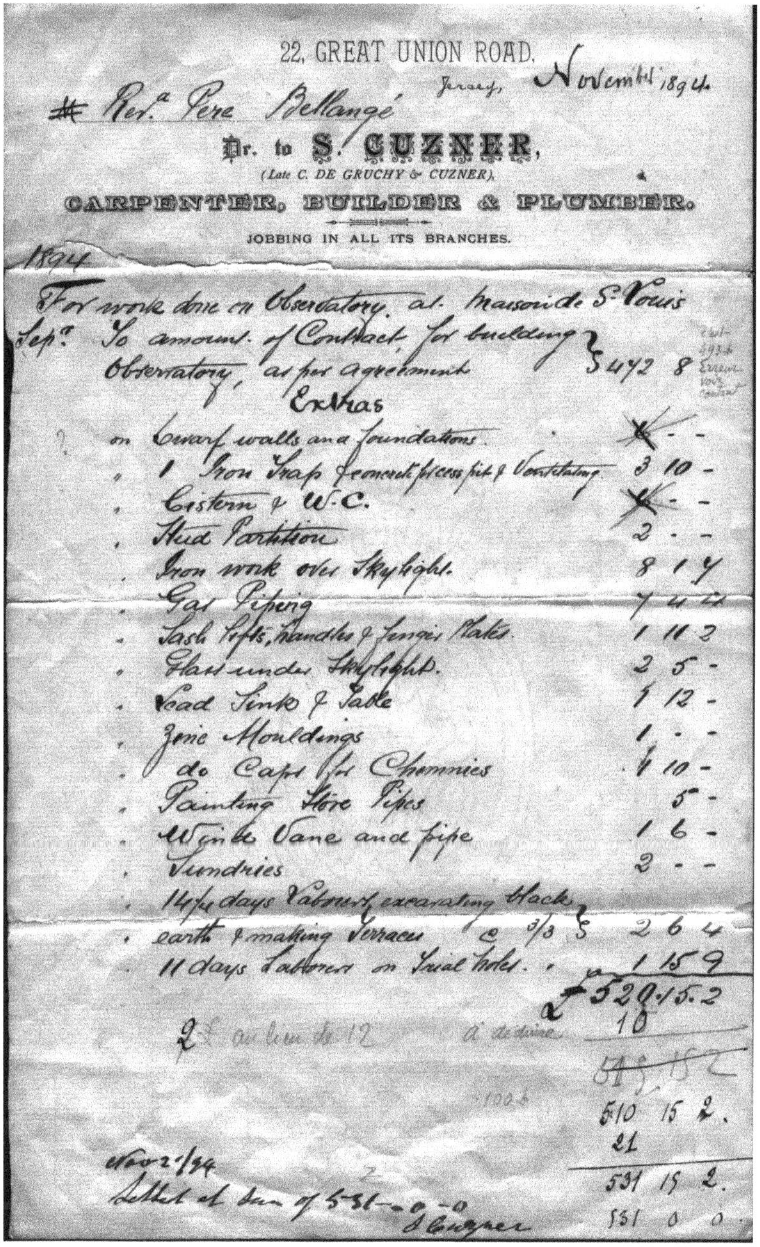
22, GREAT UNION ROAD, Jersey, November 1894.

Revd Pere Bellangé

Dr. to S. CUZNER,

(Late C. DE GRUCHY & CUZNER).

CARPENTER, BUILDER & PLUMBER.

JOBBING IN ALL ITS BRANCHES.

1894

For work done on Observatory at Maison de St Louis

Sept. To amount of Contract for building Observatory, as per agreement 492 8

Extras

on Dwarf walls and foundations

" 1 Iron Trap & cement pieces fit & Ventilating 3 10 -

" Cistern & W.C.

" Stud Partition 2 - -

" Iron work over Skylight 8 1 7

" Gas Piping 7 4 4

" Sash lifts, handles & finger plates 1 11 2

" Glass under Skylight 2 5 -

" Lead Sink & Table 1 12 -

" Zinc Mouldings 1 - -

" do Caps for Chimnies 1 10 -

" Painting Stove Pipes 5 -

" Wind Vane and pipe 1 6 -

" Sundries 2 - -

" 14½ days Labourer excavating back earth & making Terraces @ 3/3 2 6 4

" 11 days Labourer on Trial holes 1 15 9

£520.15.2

2£ au lieu de 12 à déduire 10

510 15 2.

21

531 15 2.

531 0 0

Nov 2/94

Settled at sum of £531-0-0

S Cuzner

Figure 10: Builder Mr Samuel Cuzner's account for £531 for the construction of St Louis Observatory in 1894. (St Louis Observatory Archive)

Figure 11: A view of the newly built Maison St Louis Observatory and adjacent tower. This undated image was probably taken soon after the anemometer was fixed in place in February 1895. The anemometer was 112 metres (367 feet) above mean sea level and reached by 254 steps. (St Louis Observatory Archive)

Figure 12: Dechevrens specially designed *Anémomètre Universel* capable of measuring horizontal and vertical components of the wind. (St Louis Observatory Archive)

were added (a similar barograph remains in working condition) and in June 1897 he started recording terrestrial magnetism, which continued for ten years.[18] In 1898 a Jordan sunshine recorder came into use, then a Besson comb nephoscope, for which he recommended a modification, was added for cloud observations in 1903. In 1910 it was a recording rain gauge, then a radio antenna was fixed to the tower in 1913. The antenna was connected to a ceraunograph, a radio device for counting lightning discharges, though there is no further mention of this equipment.[19]

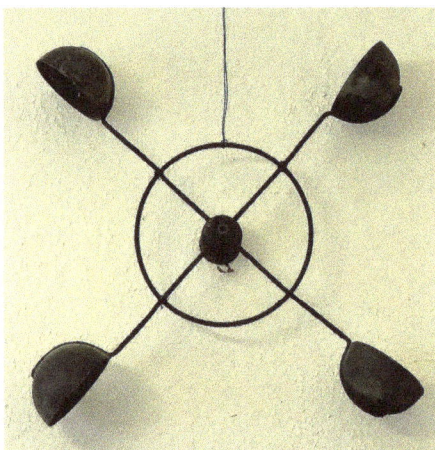

Figures 13-14 (above): Surviving parts of Dechevrens' *Anémomètre Universel* originally on the top of the tower.

Figure 15: Anemometer cups originally in use on the roof of the Observatory.

Figure 16: The wind direction vane which can be seen on the roof in early photographs of the Observatory.

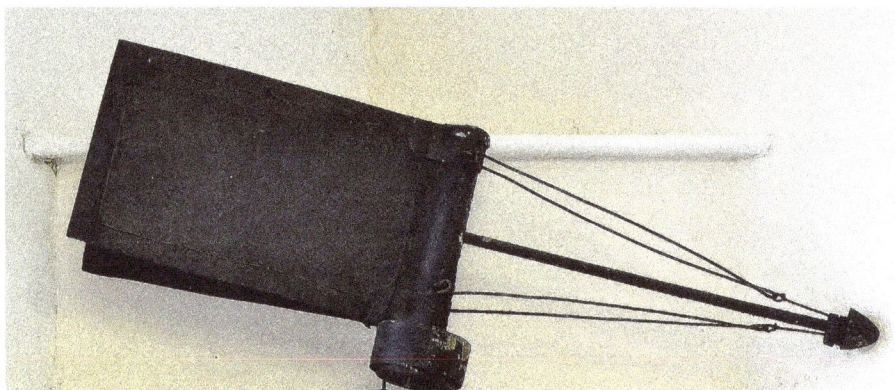

Figure 17: The Besson comb nephoscope was an instrument for measuring direction of movement and the speed of clouds. It is seen here in front of the Observatory, with Fort Regent in the background. (St Louis Observatory archive)

Figure 18: A general view from the front door of the Observatory showing from left the Besson nephoscope, a rain-gauge and the louvred thermometer screen. In the background are Victoria College, Fort Regent and Elizabeth Castle faintly on the right. (St Louis Observatory archive)

Vue prise du seuil de l'Observatoire
La Herse néphoscopique   Le Pluviomètre   L'Abri des thermomètres

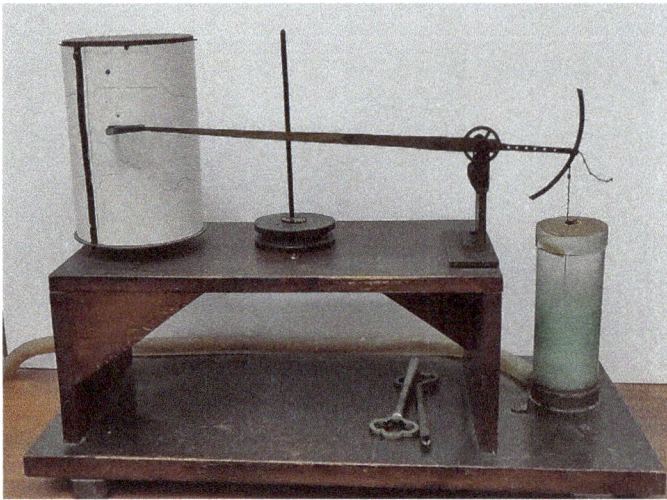

Figure 19: An air and water microbarograph still in working order at the Observatory. It is believed this example was made by Father Rey in 1917, but it would have been very similar to Dechevrens' earlier version. (Author's photograph)

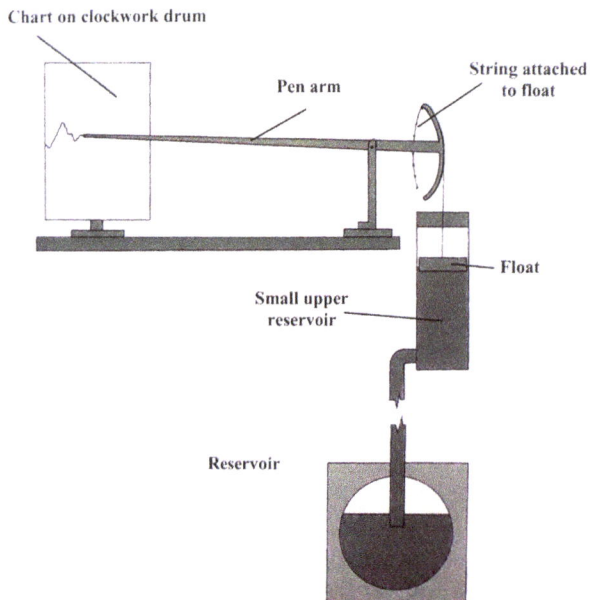

Figure 20: A diagram explaining how the air and water microbarograph works. The small reservoir and chart in the central room of the Observatory are connected by a tube to the main reservoir in the basement, which is encased in concrete to minimise temperature changes. (Jersey Meteorological Department)

Figure 21: Dechevrens used the well-known French instrument makers *Richard Frères* to build some of his instruments. He also bought instruments from the company such as this fine example of a hand held anemometer. (Author's photograph)

Figure 22: An impressive letterhead on this account from instrument makers *Richard Frères* for repairs to a weathervane in 1896. (St Louis Observatory Archive)

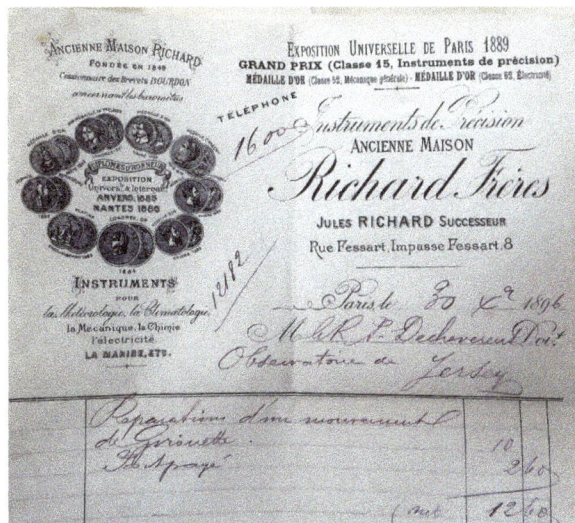

We do not know how much of the work Dechevrens carried out on his own, but it is said that he climbed the tower regularly. As another of those who did so while a young student meteorologist, Charles Rey attested that to climb 250 steps in all weathers was no easy task (Rey 1960). In the Observatory's daily routine he would have been assisted by young Jesuits training for work in the overseas observatories. Indeed it would have been impractical to maintain the routine of taking eight weather observations each day without help. We know that Père Léon Des Hayes acted as his assistant for some years from 1900 and the 1911 Census helpfully identifies two more men 'attached to the Observatory'. Thus, in its heyday four men would have operated some form of shift work.[20]

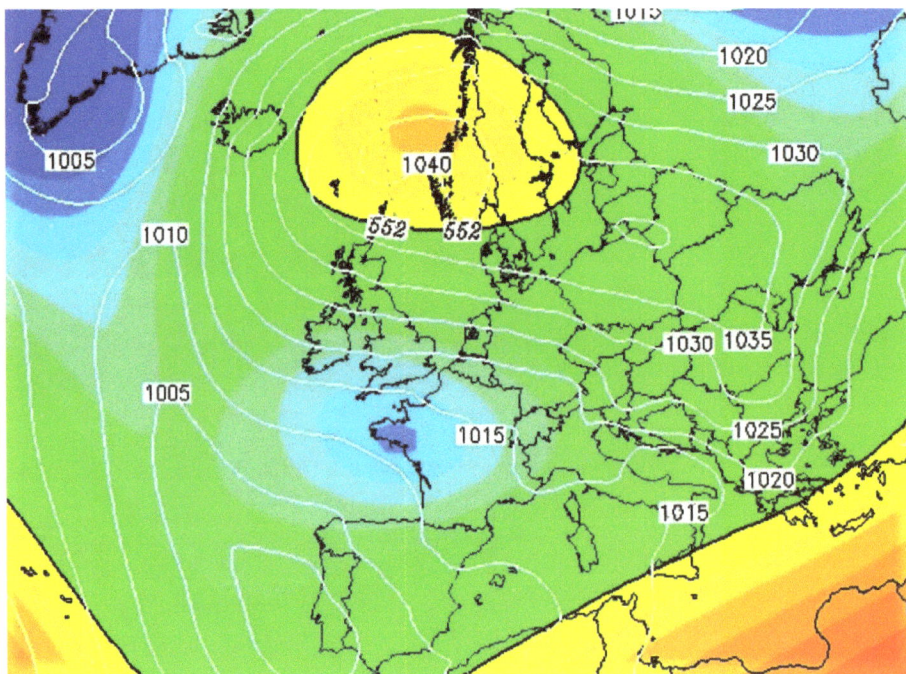

Figure 23: Pressure chart for 5 January 1894, when five days after opening the coldest temperature recorded at the Observatory was read. The darker blue area over Brittany is the centre of a pool of very cold air extending high into the atmosphere. (Chart from wetterzentrale.de, using data from NOAA's 1.0° reanalysis)

The climatological records show that the early days of the Observatory were distinctly cold. A minimum temperature of -10.3°C was read by Dechevrens on 5 January 1894, the fifth day of operation, and remains in late 2021 the lowest temperature registered at the station.[21] In the following winter the mean temperature of -1.0°C in February 1895 is still 127 years later the only month to register a sub-zero mean temperature. Also in Dechevrens' time the mercury touched 1051.7 millibars on the morning of 29 January 1905, Jersey's highest pressure reading and the 266.4 millimetres of rainfall in November 1910 remains the highest recorded in any November.

Publications

The annual series of *'Bulletin des Observations'* compiled by Dechevrens provided a comprehensive annual summary of the St. Louis Observatory's weather and climate, which form the basis of the Island's climatology today. The reports include detailed comparisons between winds near the ground and those recorded by his *'anémomètre-universal'* at the top of the tower, also calculations of diurnal temperature and pressure variations, with their mathematical harmonics. A handwritten bibliography with notes by Dechevrens (c. 1920) and prefaces in the annual bulletins make it clear that he continued actively with research in Jersey, particularly his study of the vertical temperature structure of the atmosphere, upper air currents and vertical currents in cyclones, which he started at Zi-ka-weï. He published at least nine papers on these subjects between 1888 and 1908.

As was the case in China, Dechevrens was not afraid to take issue with those leading figures of the day, who used or disputed his work and its conclusions. He was aware that at best his discoveries were not being properly credited to him and at worst simply ignored and accredited to others. In a handwritten manuscript (undated c. 1912) he lists his important publications with a brief summary of their contents. Alongside he comments as to how they were received by other researchers and it is here that we can sense his frustration. He named several leading figures who used his results or mentioned them, but without acknowledging his work on the atmosphere's temperature structure, including W. M. Davis and Henry Hazen (both USA), Arthur Wagner (Austria), W. H. Dines and Charles Cave (both UK) and Jean Vincent (Belgium). They mostly accredited his discoveries to Austrian Professor Hann, with whom he appears particularly angry. As early as 1900, in a note to the *Société Météorologique de France*, read at their meeting on 9 November (p.6), Dechevrens reminded them that he had shown temperature decreased rapidly with height in areas of low pressure, before Hann had done so.[22] Again, in the Observatory's 1905 *'Bulletin des Observations'* (pp. 1-7) in a reply to Hann, he wrote a strongly worded reminder of his 20 years of research and his astonishment that Hann's treatise *'Lehrbuch der Meteorologie'* failed to acknowledge or even mention his work on the atmosphere's vertical temperature structure:

> *Je m'étais étonné . . . aucune reference à mes recherches et publications de 1886, 1887, 1898, 1902 et 1905).*
> *I'm astonished . . . no reference at all to my research and publications in 1886, 1887, 1898, 1902 et 1905).*

Hann had the advantage of being a professor and editor of an influential journal. Sadly, despite his best efforts on this important issue, Dechevrens was effectively ignored and side-lined, never to receive the credit he deserved.

Many of his papers were published in journals of leading scientific societies in France, Italy, the United States, Belgium and England. Aside from meteorological theory, he discussed a wide range of topics, ranging from the climate and sunshine in Jersey, to bolides (meteors which explode in the atmosphere), electrical currents in the atmosphere, as well as comments on synoptic weather features such as fog, high pressure, cold spells and inversions. Some were published independently and in 1900 he read a paper at the *Congrès météorologique de Paris*. His research concerning the wind climatology of Jersey was published by the French weather service shortly before his death (Dechevrens 1923). An extensive bibliography of his known papers can be found in the Appendix.

Other interests

Despite his dedication to meteorology, Marc Dechevrens had at least one other interest. Being a largely a self-taught mathematician, may explain why in the mid 1890s he designed and constructed a 'campylograph', a device for drawing geometric curves (derived from the Greek word *campylos* meaning curve). It was, he wrote, conceived and constructed completely aside from his meteorological interests (Dechevrens 1920). The device is a type of harmonograph, which in its simplest form consists of two coupled pendula oscillating at right angles.[23] A pen or pencil is attached to one and a writing surface to the other, with the gyratory motions producing intricate geometric patterns. Dechevrens' machines were more sophisticated, with his original version said to be the first to

Figure 24: An illustration of Dechevrens' *'grande forme'* campylograph. (St Louis Observatory archive, originally published in the journal *La Nature*)

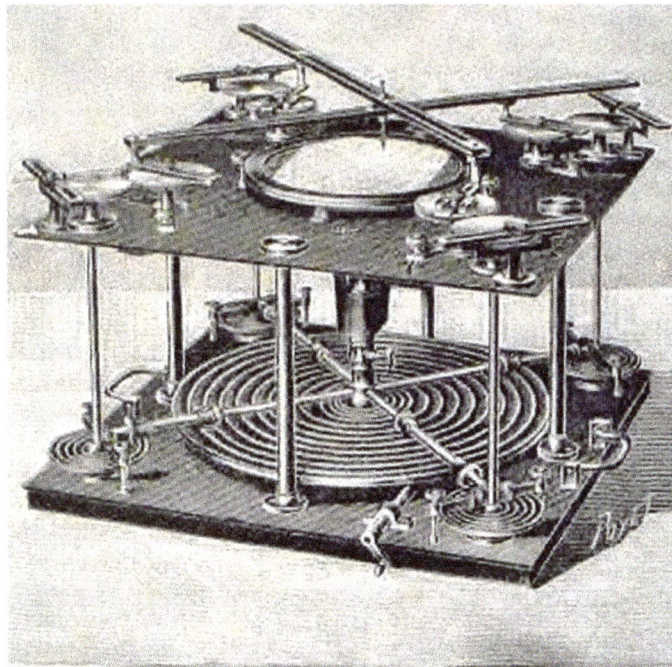

Fig. 1. — Le Campylographe du P. Marc Dechevrens, directeur de l'Observatoire de Jersey.

draw what are known in mathematics as Lissajous curves.

The second model which he termed '*grande forme*' was more complicated. Up to five movements were possible and Whitaker (2001) states it was the most elaborate such harmonograph built. The apparatus produced a graphical solution to a range of algebraic equations, creating curves which Dechevrens justifiably claimed 'leave the world of mathematics for the world of art'. He also described how by applying a small offset to the drawings a stereoscope could be used to view the curves in three dimensions (Dechevrens, 1900). Clearly his campylograph was a sophisticated piece of apparatus and one can but imagine what he might have achieved with the use of a modern computer.

In a paper presented at a meeting of the *Société Française de Physique* in June 1894, Dechevrens described his first device in detail and wrote three more descriptive papers about it. French mathematician François Apéry (2012 personal communication) also noted that it could be used to track the orbit of Venus and suggested this might explain why he designed the device. An advertising flyer from the company '*Château Père et Fils*' of Rue Montmartre, Paris in 1900 clearly indicates the apparatus was for sale, but the price was omitted. It is believed that at least one example went to an Indian prince and a model exists today at the Henri Poincaré Institute in Paris.[24] It is no surprise to read in Whitaker's paper that Dechevrens corresponded with William Rigge of Creighton University, Omaha, USA, because Rigge was the expert of the day in the rather niche world of harmonographs and he too was a Jesuit. Rigge (1926) wrote that Dechevrens was preparing a book of his campylograph designs and indeed some examples found at the Observatory are numbered, but it appears he died before this was achieved.[25]

The later years
Due to an unspecified health problem Dechevrens left for France in September 1909 and returned after recuperation in August 1911.[26] In his absence the Observatory continued to function normally with Louis Fougérat's name appearing on weekly reports and in the annual bulletins.

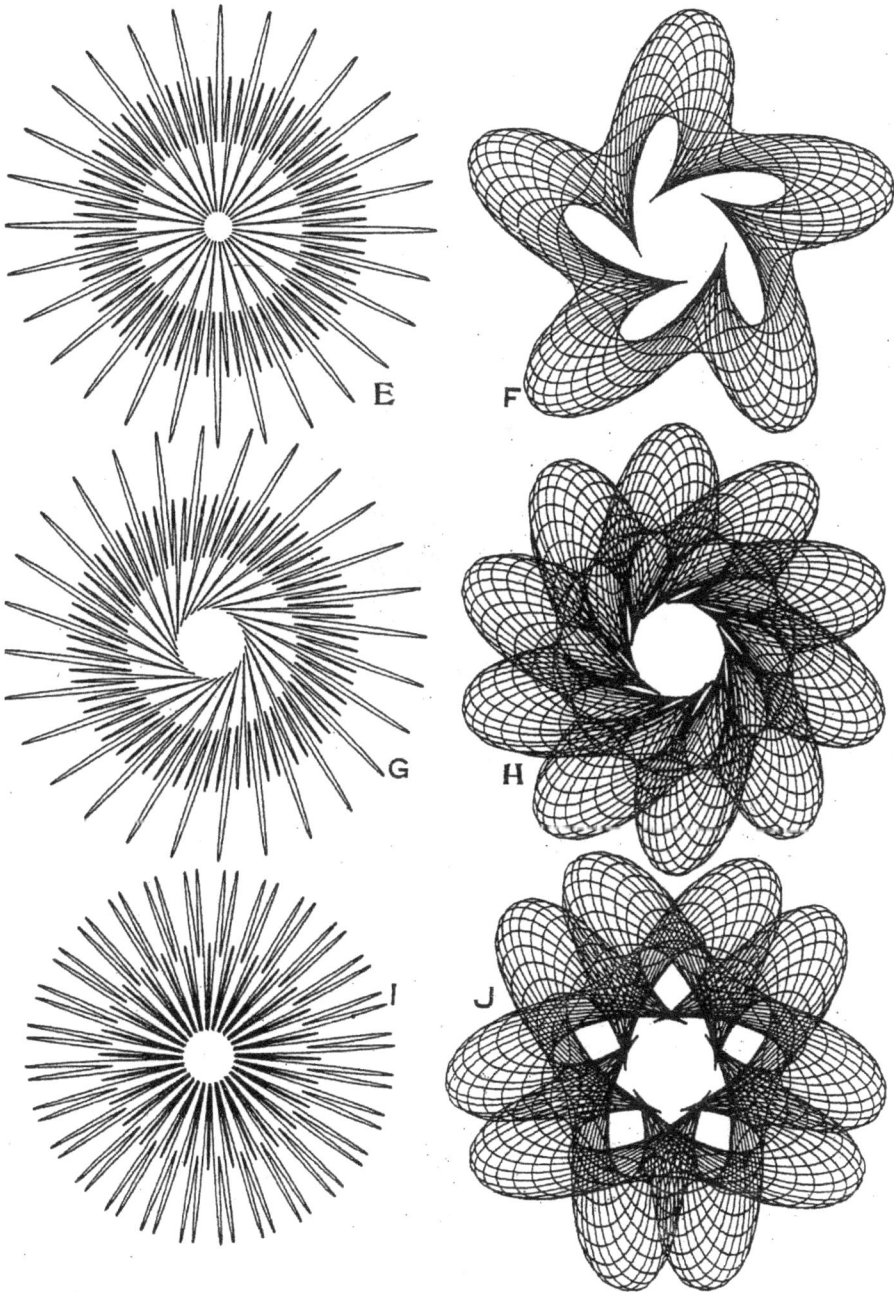

Figure 25: Captioned 'Devices for Obtaining Beauty'. Campylograph patterns leaving the world of mathematics for the world of art. (St Louis Observatory archive)

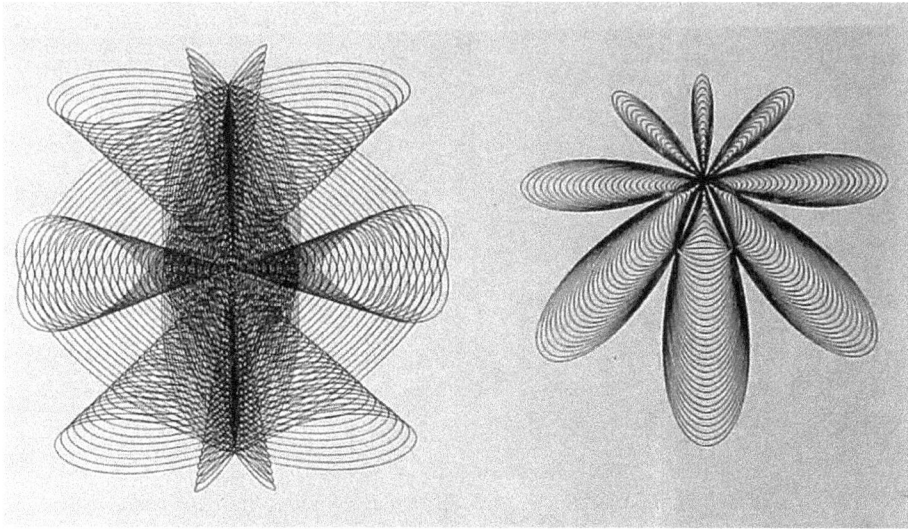

Figure 26: Intricate campylograph patterns. (St Louis Observatory archive)

However, several years later, the outbreak of war in Europe in 1914 caused considerable difficulties for the Observatory. About seventy students and teachers, over half the total complement, suddenly left the Island to return to their home countries or join the military forces. Among them, twenty-nine French nationals answered the call to mobilise, of whom eleven paid the ultimate price and fell on the battlefield.[27]

In a short note accompanying a one-page summary of the 1914 readings, Dechevrens implied that the end of the Observatory was imminent.

Après 21 années d'existence l'Observatoire St Louis de Jersey est contraint de disparaître devant des difficultés . . . Son directeur et fondateur remercie vivement les nombreux correspondants . . .

However, he vowed to continue with some recordings and to study data he had accumulated over the previous 20 years. With diminished resources and a severely depleted St Louis community, he reduced weather observations to thrice daily, compared with the previous eight. Aged nearly 70 years at the outbreak of World War I and without help to maintain the instruments on the tower, wind recordings from the summit ceased at this time. The comprehensive annual *'Bulletin des Observations'* was another casualty, but a short annual résumé was published in the UK Met Office's Geophysical Memoirs and also in the annual bulletin of the

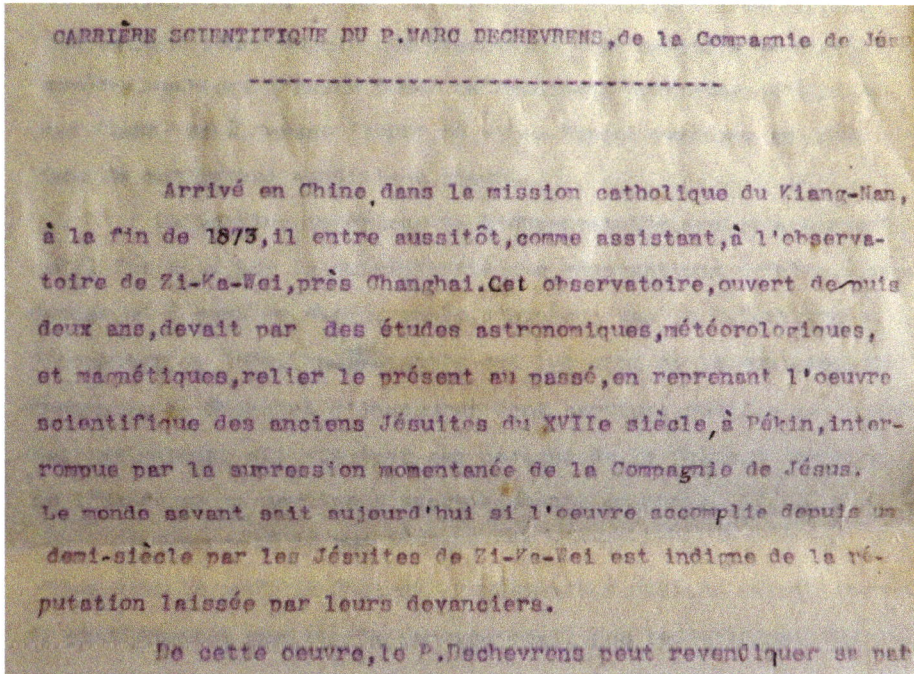

Figure 27: Dechevrens' career summary typescript which runs to six pages.

Figure 28: Part of Dechevrens' summary of his published papers, written in his own hand. From the text it was written in 1920, the year he ceased taking weather observations.

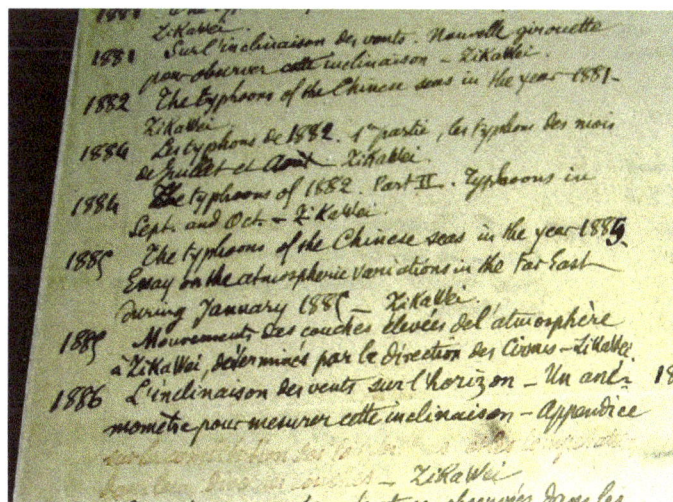

Société Jersiaise, where it appears to this day.

On a positive note, the war years gave our priest-meteorologist more time to explore several other fields in parallel with meteorology, as mentioned in his career summary (Dechevrens 1920). These were first, the earth's magnetic currents, second telluric currents (underground and undersea electrical currents), third the effect of the oceanic tide on the currents particular to Jersey, and lastly a study of the air's potential electric currents. Of the latter he mentions that a first series of studies had been made in 1889 with a portable electroscope, implying his presence in Jersey at that time. Nearly all his published work from 1914 onward explored these subjects, though they are not related to meteorology.

At the end of 1920 Père Dechevrens finally called a halt to his 27 years of Jersey weather recordings, but he soldiered on with a few magnetic readings for another three years until shortly before his death. It was on 6 December 1923, aged 79 years, that he breathed his last at Maison St Louis, succumbing to heart disease.[28] Coincidentally it was within a few days of the 50th anniversary of his arrival at the Zi-ka-weï Observatory.

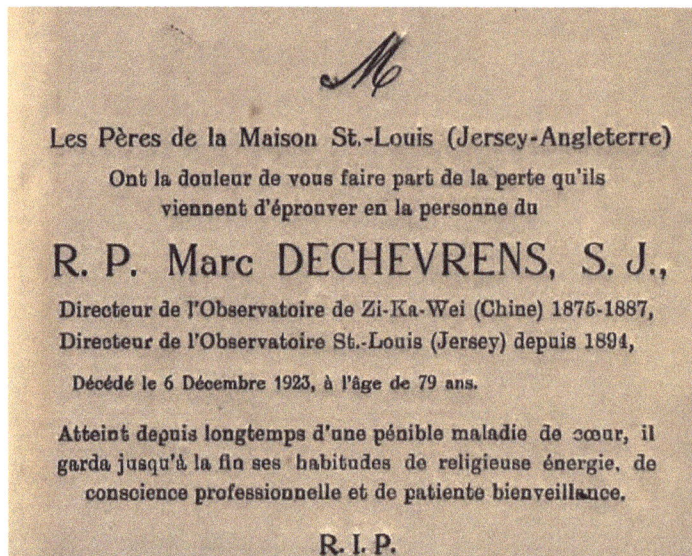

Les Pères de la Maison St.-Louis (Jersey-Angleterre)

Ont la douleur de vous faire part de la perte qu'ils
viennent d'éprouver en la personne du

R. P. Marc DECHEVRENS, S. J.,

Directeur de l'Observatoire de Zi-Ka-Wei (Chine) 1875-1887,
Directeur de l'Observatoire St.-Louis (Jersey) depuis 1894,

Décédé le 6 Décembre 1923, à l'âge de 79 ans.

Atteint depuis longtemps d'une pénible maladie de cœur, il
garda jusqu'à la fin ses habitudes de religieuse énergie, de
conscience professionnelle et de patiente bienveillance.

R. I. P.

Figure 29: The announcement of Père Marc Dechevrens' death by the Fathers at Maison St Louis in December 1923. (Author's photograph)

Figure 30: Marc Dechevrens' legacy. The Maison St Louis Observatory seen here in summer 2021. It continues to record the weather and is recognised as a Centennial Station by the World Meteorological Organisation.

Maison St Louis Observatory – Dechevrens' legacy

Fortunately, after four years of near inactivity, the Observatory was revived soon after Dechevrens' death with the arrival of Père Christian Burdo (1881-1961) in 1924. The building took on a new lease of life as he took up its cause helped by Père René de Vallois. The instruments began working again, with thrice daily recordings from 1 January 1925 and since that date not a single day is missing, with the Observatory providing Jersey's only weather record for the German Occupation years from 1940 to 1945. Burdo was succeeded in 1934 by Père Charles Rey (1897-1982), until he retired in 1980 following an accident and since then the Observatory has been operated by the Jersey Meteorological Department (JMD), a branch of government.[29] For several decades it has acted as the reference station for Jersey's climate, the solid foundation of which is due to Dechevrens' diligent recording in the first 27 years. In September 2020 the JMD learned that an application for Centennial Station status had been approved by the World Meteorological Organisation in recognition of its long series of weather recordings at the same site. It is recognition of which Marc Dechevrens would surely have been proud and is a fitting legacy to the work he started there.

Summary

Of the man himself we know little, except that according to an obituary in the *Patrie Suisse* newspaper Marc Dechevrens had a lively, sparkling character, but his career was a long list of achievements. On the practical side he designed and built instruments, particularly to take wind measurements, he established a time service in Shanghai and pioneered a typhoon warning service along the China coast, to the great benefit of mariners and the Chinese community in general. Under his leadership he developed the Zi-ka-weï Observatory into a first rate institution with an excellent reputation in the Far East. It continued operating under distinguished successors, Louis Froc, Pierre Lejay and Ernesto Gherzi until the Jesuits were expelled in the communist takeover in 1949. This was all very much in keeping with the Jesuit principle of engagement with the world and contributing to the improvement and progress of society. In Jersey, Dechevrens' outstanding achievement was founding the Maison St. Louis Observatory. He guided it through the first 27 years and as a lasting legacy the weather is still recorded there today, adding to Jersey's climatology nearly 100 years after his death.

In parallel with his practical achievements Dechevrens' research can be found in many papers, notes and memoranda. Within this contribution to international scientific dialogue, several of his papers made important, if not pioneering, contributions to meteorological theory. This was particularly so in respect of the vertical temperature structure of cyclones and movement of air currents in the upper atmosphere. Unfortunately, others took the credit for some of this work, a fact of which he was well aware, but his attempts to set the record straight were largely unsuccessful.

Conclusion

In recent decades Jesuit priorities have changed and no longer focus on science. In parallel, most countries have developed their own national weather services and station networks, so the *raison d'être* for Jesuit observatories ceased some time ago. Since the 1950s therefore, many of the once busy observatories have closed while others, like the Maison St Louis Observatory in Jersey, have been integrated into national weather services or merged with other scientific bodies. However, it would be remiss of historians to forget important contributions to the development of meteorology and other sciences made by the Jesuit scientists who worked in the observatories and in particular the dedication of individuals such as Marc Dechevrens.

Acknowledgements

My thanks to Professor Robert J. Whitaker of Missouri State University who alerted me to the world of harmonographs and Professor François Apéry for additional information about the campylograph. Also Gaston Demarée of the Royal Meteorological Institute of Belgium for sending images of the Zi-ka-weï Observatory and Chinese student Yang Li who while at Nottingham University researched the Chinese script around Père Dechevrens' portrait. It is also a pleasure to acknowledge the support of my former employers at the Jersey Meteorological Department by allowing access to the Maison St Louis Observatory and agreeing to my use of archive material.

Notes

1   Zi-ka-weï was the spelling used by the Jesuits while at Shanghai and has been adopted in this work. In the 19th century the name was sometimes unhyphenated, at other times spelt Ziccawei or Siccawei. In the 19th century Zi-ka-weï was a village several kilometres to the southwest of Shanghai, but now lies in the busy commercial area of Xujiahui, within the Xuhui district of a hugely expanded city. It is believed the original observatory building was replaced in 1906. This replacement building still exists and was visited by the author in 1995, who was told it was used as offices. In late 2012, the China Meteorological Administration (CMA 2012, see also WMO 2012) celebrated the 140th anniversary of the 'Shanghai Xujiahui Observatory'. The press release contains correct historical details, except that the Jesuit-French connection is almost entirely expunged. It also states that the Observatory has been formally recognized as a Centennial Climate Station by WMO, but it does not appear on the latest map of those stations (WMO 2021).

2   Weather recordings for 1893 were found at the Observatory signed by Jesuit Père Charles Noury, a physics teacher at Maison St Louis. He was also a noted geologist, having written the first comprehensive study of Jersey geology (*Géologie de Jersey* 1886) and was elected a Membre d'Honneur de la Société Jersiaise in 1887. His weather recordings survive as monthly summaries not daily records and lack detail of the site, instruments and instrument exposure. For this reason the Maison St Louis record is deemed to start on 1 January 1894.

3   English Jesuit Stephen Perry FRS (1833-1889), was one of the foremost scientists of his day. As an astronomer he was particularly interested in solar physics and led several expeditions to view Venus transits and solar eclipses. He died on his last one, having observed an eclipse in French

Guiana. He greatly developed the meteorological work at Stonyhurst College Observatory, as its director from 1860 to 1863 and from 1868 until his death in 1889. In 1866 it was chosen as one of only seven first order stations in the UK, when the Meteorological Office came under the auspices of the Royal Society in the 1860s. Under Perry, Stonyhurst became a center for training Jesuits who were destined to became directors of their observatories. It was at this station that Dechevrens learnt the routine of weather observing and recording before departing for China. Perry also engaged in research of the Earth's magnetic field and with Jesuit William Sidgreaves, undertook important magnetic surveys in France and Belgium in the late 1860s, but he is another largely forgotten scientist. It was almost certainly from Perry that Dechevrens acquired a similar interest, which he continued to the end of his life, though the Earth's magnetic field is not a branch of meteorology.

4   A distinction can be made between mid-latitude and tropical storms. Both have cyclonic rotation and in the right atmospheric conditions can produce very strong winds which pose great danger to shipping. However, in mid-latitudes the storms have fronts separating air masses and are associated with the jetstream, typically moving forward at speeds of 25 knots or more. Tropical storms do not have fronts and typically move at less than 20 knots.

5   The Beaufort wind scale was devised for sail-carrying vessels by Francis (later Admiral) Beaufort in the early 19th century. It was officially adopted by the Royal Navy in the 1830s and soon became widely used. Luke Howard was an English amateur meteorologist who in 1803 published a cloud classification system. It also came to be widely used and the system in current use is a development of his original. English colonial officer and ex-mariner Henry Piddington (1797-1858) was based in Calcutta from 1830. With a general interest in science, from 1838 he studied storms, inspired by William Reid's 'Law of Storms' book. From 1840 until his death, he wrote 20 memoirs, published in the Journal of the Asiatic

Society of Bengal. Available online, the memoirs make fascinating reading and show that Piddington had a profound understanding of storms from the point of view of seamanship at this early stage in the science of meteorology. In 1844 he published a book *'The Sailor's Hornbook for the Law of Storms'*. In the second edition in 1848, he is credited with being the first person to coin the word 'cyclone' to describe any 'turning gale or tyfoon'.

6   A report of this meeting is available online: <https://en.wikisource.org/wiki/First_International_Maritime_Conference_Held_for_Devising_an_Uniform_System_of_Meteorological_Observations_at_Sea>

7   The typed manuscript is undated and unsigned, but the text makes clear that it was written in 1920. For example on page 4 we read *'aujourd'hui l'observatoire de Jersey compte 26 années d'éxistence'* - it opened in 1894. Though written in the third person, it contains corrections in the distinctive handwriting of Dechevrens and a complicated explanation of the design and working of the campylograph (see later) can only have come from its inventor. For this reason the typescript has been designated as Dechevrens, (1920).

8   The Secchi meteorograph was an early form of automatic weather station, named after its Italian inventor Jesuit Angelo Secchi (1818-1878) and sometimes called a meteograph or meteorgraph. Secchi stayed at Stonyhurst College for a time in 1849 and while in America met influential oceanographer Matthew Fontaine Maury, with whom he corresponded for many years. From 1849 until his death he was Director of the Roman College Observatory. He had wide-ranging scientific interests being a pioneer in the field of astrophysics, especially stellar spectroscopy and one of the first to state with authority that the Sun is a star. Another long lasting legacy was developing a technique for measuring the clarity of sea water which is still used today. He skilfully developed his meteorograph over ten years from 1857. Using electro-mechanical relays, it continuously

recorded pressure, wind speed and direction, temperature, rainfall, and humidity. In early form it was relatively simple, but became a complex piece of machinery by the time it was shown and admired at the 1867 Paris Exhibition. Due to the high cost and complexity it did not find the widespread use Secchi had hoped for (Brenni 1995), but examples were made for the Manila, Zi-ka-weï, Belén, Washington and Palermo observatories. An example exists in the modern era at a museum at Monte Porzio, near Frascati, Italy, with possibly another in the Philippines.

9   When a tropical storm reaches a mean wind speed of 64 knots (74mph) or more it is called a typhoon. They are purely tropical phenomena and immensely destructive with the mean wind speed in super typhoons blowing at 160 knots (184 mph) or more. On occasion they move into mid latitudes, usually in a dissipating phase, becoming extra-tropical but can remain very potent. Hurricanes (in the Atlantic Ocean and eastern Pacific Ocean), tropical cyclones (in the Indian Ocean) and typhoons (in the west Pacific and China Seas) are the same phenomena with different names.

10   Benito Viñes (1837-1893), a Spanish born priest-meteorologist, was Director of the Belén Observatory in Havana from 1870 until an untimely death in 1893. Pioneering the study of tropical meteorology, he became the foremost authority on tropical storms of his day, inspiring fellow Jesuits Marc Dechevrens (1845-1923) in Shanghai and Francisco Faura (1840-1897) in Manila. Faura corresponded with Viñes and realised that hurricanes and typhoons were the same. Viñes developed a Caribbean weather reporting network and with many successful forecasts from 1875 onward, made life safer for the local population. The interesting story of his life, using original sources, is available in a book by Ramos Guadalupe (2014).

11   Dr William Doberck (1852-1941), Director of the Hong Kong Observatory from 1884 to 1908 was highly regarded as an astronomer,

but did not fully embrace meteorology. He was an irascible character at odds with the Colonial Office, the Hong Kong Governor, the Chamber of Commerce, local marine interests and even his own staff. For Doberck tact and compromise were alien concepts. The Jesuits were well trained and neither ignorant nor arrogant, but he held a particular dislike for their well organised observatories at Zi-ka-weï and Manila, regarding their existence as rivalry and an 'intolerable affront'. He wanted to suppress the 'superfluous observers' (MacKeown 2010 p.46) because they thwarted his ambition to become top dog on the China Coast. The querulous Doberck simply refused to co-operate, but astonishingly survived in his post for 24 years. MacKeown (2010) provides much detail on Doberck's character and the background to Hong Kong Observatory.

12   Swedish meteorologist H. H. Hildebrandsson (1838-1925) was Professor of Meteorology at Uppsala University. He was particularly noted for his work on clouds and was the first person to use photography in the study and classification of cloud forms. With others, he produced the first edition of the International Cloud Atlas in 1896.

13   Austrian meteorologist Julius von Hann (1839-1921) made important contributions to meteorology and climatology. He headed the Austrian national weather service for a time and held professorships at Graz and Vienna. He was also editor of the influential journal *Meteorologische Zeitschrift* for more than half a century (1866 to 1920).

14   Frenchman Léon Teisserenc de Bort (1855-1913) was for a time head meteorologist of the French weather service and a pioneer in the field of aerology. Using unmanned sounding balloons, the forerunners to modern radiosondes, he investigated the upper atmosphere. Jointly, with Richard Assman from Germany, he found that above about 11 kilometres air temperature ceased to fall with height and became isothermal. He named this newly discovered layer the stratosphere.

15   Aliens Registration card, Jersey Archive D/S/B1/743

16   Rey (1960) provided details of the design and construction of the Jesuit tower. It was built for £1,260 by Monsieur T. Seyrig of the '*Société Anonyme de Construction et des Ateliers*' of Willebroeck in Belgium, who had a reputation as a first class engineer. A contract was signed on 6 April 1894 with groundworks starting on 16 May. On 18 September the first metal beam was fixed in place and by 3 November the 37-ton tower was complete. A series of photographs at the Observatory show the stages of construction over this six week period. The masonry work was undertaken by builder Mr Samuel Cuzner, of Great Union Road, who constructed the observatory. Maintenance of the tower was an expensive and arduous job, sometimes undertaken by younger Jesuits (Rey 1960), but the task finally proved too much. It was offered to the States of Jersey who declined the offer and on 20 February 1929, with snow lying on the ground, the Jesuit's tower was demolished after almost 35 years.

17   Dechevrens first made a special windvane at Zi-ka-weï in 1879, to show the inclination of the wind with respect to a horizontal plane. In 1881, he substituted it with his 'clino-anemometer' which could also measure the vertical component of the wind (Dechevrens 1920). He further developed the design to combine it with a standard direction indicator which became his '*grand anémométre-universel*'. The instrument was built by *Richard Frères* in Paris and placed atop the St Louis tower some 112 metres (367 feet) above sea level in February 1895, but in practice it was too complicated for widespread use.

18   The advantage of an air and water microbarograph lies in its sensitivity. As water is far less dense than mercury, such a barograph magnifies changes in air pressure by a factor of five and a half compared with a mercury barometer, which can be important in research work.

19   Ceraunographs are radio devices which detect and record the electrical discharges in storms, which are seen as lightning and heard as radio 'static'. We should not be surprised that Jesuit scientists were involved in their development. In 1900 at Kalosca Observatory in Hungary, Jesuit astronomer Gyula Fényi (1847-1927) and his assistant Johann Schreiber conceived of a ceraunograph and following experiments in 1901 it was subsequently offered for sale. Skilled at mechanics as well as science, in the same year Jesuit meteorologist Frederick Odenbach (1857-1933), Director of the observatory and Professor of astronomy and meteorology at St Ignatius College, Cleveland, USA made his own ceraunograph. His aim was to warn of the arrival of thunderstorms. A description and photograph of Odenbach's instrument, can be found in a St Louis University (1912) publication. It was Odenbach who gave the device its now accepted name of 'ceraunograph'. Mention should also be made of Frenchman Albert Turpain (1867-1952), a wireless pioneer (but not a Jesuit) who made storm-detecting experiments in 1902. He wanted to improve thunderstorm forecasts to optimize the use of hail cannons in wine growing areas. Dechevrens does not say which ceraunograph type was installed at the Observatory, but about 25 of Turpain's instruments were manufactured by *Richard Frères* in Paris. It was the same company that built Dechevrens' anemometer some years earlier and from whom he bought other instruments. A history of the development of ceraunographs is covered in two papers by Brenni (2020 and 2020a) who concluded that they caused over-optimism regarding the possibility of predicting thunderstorms which led to their demise after World War I.

20   In addition to Dechevrens, Léon Deshayes (sic) 72 years, Henri Defert 48 years and Xavier Comyers 42 years, are listed in the 1911 Jersey census as 'attached to the Observatory'. Unfortunately the 1901 census did not contain such detail.

21   It is not the lowest temperature recorded in Jersey. On 18 January 1891 official recorder John Fisher measured 12.2°F (-11.0°C) at St Aubin and

on 20 January 1963 the temperature at Jersey Airport fell to -10.9°C (with -9.6°C at the Observatory that night).

22   *Le R[everend]. P[ère] Dechevrens a rappel dans une note très substantielle qu'il avait montré avant M. Hann la rapide décroissance des températures dans les zones de basse pression et en opposition avec ce caractère les températures relativement élevées qu'on trouve dans les maxima de pression.*

23   By coincidence, the author used a basic harmonograph in art lessons at Hautlieu School, Jersey in the 1960s, about 200 metres from where Dechevrens designed his device about 70 years earlier. Harmonographs of varying complexity have been invented over the years and when Professor Robert Whitaker of Southwest Missouri State University contacted the JMD, it alerted the author to the world of harmonographs. His paper (Whitaker 2001) has a section on the Dechevrens campylograph and notes his extensive correspondence with William F. Rigge (1857-1927) of Creighton University, USA. It is another example of the tight Jesuit network in operation, for Rigge was a fellow Jesuit and an authority at the time in the very specialist world of harmonographs. An article in the Jersey Times and British Press newspaper on 15 February 1895 describes the instrument, referring to it as a 'cinégraph'.

24   An enquiry received at the JMD from French mathematician Professor François Apéry in 2011, revealed that the Henri Poincaré Institute in Paris had two incomplete examples of a campylograph with plans to resurrect a working example. In a recent paper (Apéry 2020) describes the campylograph and includes a picture (Figure 18) of what appears to be a complete machine. A short talk on the campylograph is found at: <https://vimeo.com/205017118> and a demonstration of how the curves are produced is found here: <https://pstricks.blogspot.com/2020/07/le-campylographe-avec-pstricks-1.html>

25   Notwithstanding Rigge's comment, it appears that an album of campylograph designs was published in 1907/8. See Appendix.

26   Perhaps this was due to the troublesome heart complaint mentioned on his death notice, but to the author's knowledge it is not explicitly mentioned anywhere. Dechevrens also stated in the 1908 Observatory bulletin that deterioration of his eye sight had caused him to stop taking magnetic readings.

27   The Jersey Census provides details. In 1891 there were in total 126 men at Maison St Louis, including domestic staff, and a complement of 112 in 1901. In 1911, the census nearest the outbreak of war, the complement was 140, of whom 74 were students, nearly all aged from 20 to 31 years old.

28   Following a service at Maison St Louis he was buried in Almorah Cemetery on 8 December in the St Louis Grave. *Sinnatt and Sons Funeral Directors Account Ledger, June 30th 1922 – April 28th 1925*, L/A/41/A1/21 Jersey Archive.

29   The main Jesuit house returned to France in 1954 and the following year an attempt was made to sell the Observatory to the States of Jersey (government) but the proposal was unsuccessful. Charles Rey had been the sole Jesuit in Jersey since then and by the early 1970s the aging Director was anxious that the Observatory's long weather and climate record should continue when he left. Aware of its historical value, he suggested that the States of Jersey buy it to preserve and continue the climate record. This proposal was successful and the purchase duly took place in 1974. Rey continued as the resident observer, helped by JMD staff as required until his last observation in late 1979. A smooth transition was thus assured and it passed into the control of a national weather service, rather than close as other Jesuit observatories had.

References

Apéry, F. (2020), 'The Cabinet de mathématiques of the Henri Poincaré Institute in Paris'. Kwartalnik historii nauki i techniki (Quarterly journal of the history of science and technology – Warsaw). Vol. 65 (3): 97–108.
< DOI 10.4467/0023589XKHNT.20.021.12604>

Brenni, P. (1995), 'Le météorographe du Père A. Secchi'. *La Météorologie*, 8ème série: 116-117.

Brenni, P. (2020), 'Written by Lightning. A short story of lightning recorders: ceraunographs, electrographs, klydonographs. Part 1: From Becaria's Ceraunograph to Gergely Palatin's Detector'. *Bulletin of the Scientific Instrument Society*: 145: 2-10.

Brenni, P. (2020a), 'Written by Lightning. A short story of lightning recorders: ceraunographs, electrographs, klydonographs. Part 2: From Lancetta's electrograph to Peters's klydonograph and its improvements'. *Bulletin of the Scientific Instrument Society*: 146: 24-33.

China Meteorological Administration (2012). The 140th anniversary of the Shanghai Xujiahui Observatory. <http://www.cma.gov.cn/en/NewsReleases/MetInstruments/201211/t20121109_189806.html>

Coyne, G. V. (2015), 'The Jesuits and Galileo: Fidelity to tradition and the adventure of discovery'. *Forum Italicum*: 49 (1): 154-165.
<https://doi.org/10.1177/0014585815572767>

Dechevrens, M. (1879), *Le Typhon de 31 Juillet 1879*. Zi-ka-weï. pp.iv+33+XVII. (Available online from NOAA Library, USA).

Dechevrens, M. (1881), 'Correspondance sur la théorie des cyclones'. *La Nature*: 419: 30.

Dechevrens, M. (1885), *Mouvements des couches élevées de l'atmosphère a Zi-ka-weï déterminés par la direction de Cirrus*. Zi-ka-weï. pp. 15.

Dechevrens, M. (1886), 'On vertical winds in cyclones'. *American Meteorological Journal*: 3: 170-84. (Translation of the original: Sur les courants verticaux dans les cyclones; constitution des cyclones).

Dechevrens, M. (1887), *Sur les variations de température observées dans les cyclones.* 2nd note. Zi-ka-weï. pp.17.

Dechevrens, M. (1888), 'Quel est le sens des courants verticaux au centre des cyclones?' *Comptes Rendu – Académie des Sciences.* (Avril).

Dechevrens, M. (1900), 'Le camplylographe, machine à tracer des courbes'. *Comptes Rendus - Académie des Sciences*, 11 June 1900: 130: 1616-1620.

Dechevrens, M. (1920), 'Carrière scientifique du P. Marc Dechevrens, de la Compagnie de Jésus'. (ms, Maison St. Louis Observatory).

Dechevrens, M. (c. 1920), Catalogue des publications diverses faites par le P. Marc Dechevrens S.J. en Chine et en Europe à partir de 1874'. (ms, Maison St. Louis Observatory).

Dechevrens, M. (1923), *Étude du vent à Jersey (dans la Manche).* Mémoires de l'Office National Météorologique de France, 1 (3), pp.55.

Dehergne, J. (1976), 'Zi-ka-weï l'Observatoire des cyclones'. *La Météorologie* 4ème: 179–88.

Delattre, P. (Ed) (1953), *Les établissements des Jésuites en France depuis quatre siècles.* 5 volumes. Enghien and Wetteren (Belgium).

Gautier, H. (1924), Un maître en Physique du globe. Le Père Marc Dechevrens, S.J. *Études* (5 Février).

Hann, J. von (1914), Remarks on the nature of cyclones and anti-cyclones. *Monthly Weather Review.* 42: 612-616.

Hildebrandsson, H. H. & Teisserenc de Bort L. (1903), *Les bases de la météorologie dynamique.* Gauthier-Villars, Paris.

Kington, J. (1999), 'Meteorologist's profile – William Clement Ley'. *Weather* 54: 166-172.

Kutzbach, G. (1979), *The thermal theory of cyclones. A history of meteorological thought in the nineteenth century.* American Meteorological Society, Boston, MA. pp.xiv+255. <ISBN 9781940033808>

Labrosse, F. (1875), *The Navigation of the Pacific Ocean, China Seas etc.* (Translator: J. W. Miller), United States Hydrographic Office 58, Washington: <https://archive.org/details/navigationofpaci00labrrich/>

Le Blancq, F. W. (1994), *1894 to 1994 – One hundred years of weather recording at the Maison St Louis Observatory, Jersey.* Jersey Meteorological Department. pp.18. [Unpublished booklet].

Le Blancq, F. (2005), *Jesuit meteorologist Marc Dechevrens (1845-1923).* Jersey Meteorological Department, pp.10. [Unpublished ms].

MacKeown, P. K. (2010), *East China Coast Meteorology. The role of Hong Kong.* Hong Kong University Press. pp.289. <ISBN 978-988-8028-85-6>

Margollé, E. (1879), 'Observatoire de Zi-ka-weï'. *La Nature* 319: 82-83.

Margollé, E. (1880), 'La famine en Chine'. *La Nature* 359: 314-315.

O'Brien, C. (2014), 'Deliberate Confusions'. *History of Meteorology* 6: 17-34. <http://meteohistory.org/scholarship/history-of-meteorology/>

Ramos, Guadalupe E. (2014), *Father Benito Viñes: the 19th century life and contributions of a Cuban hurricane observer and scientist.* (Translator: Oswaldo García). American Meteorological Society, Boston, MA.

Rey, C. (1960), 'The Observatory; Maison Saint Louis'. *Annual Bulletin of La Société Jersiaise* 18 (1): 313-320

Rigge, W. F. (1926), *Harmonic Curves.* The Creighton University, Omaha, NE. pp.213. <https://archive.org/details/in.ernet.dli.2015.515833/page/n9/mode/2up>

Rochas, M. (2005), 'H. Peslin, ingénieur des Mines à Tarbes'. *La Météorologie* 8ème séries, 49: 42-45.

Rosser, W. H. (1886), *The Law of Storms considered practically.* Charles Wilson, London. 2nd Edition. pp.156. <https://archive.org/details/lawofstormsconsi00ross/page/64/mode/2up>

Stierli, J. (2005), Dechevrens, Marc, in *Dictionnaire historique de la Suisse (DHS)*, version du 26.01.2005, translated from German. <https://hls-dhs-dss.ch/fr/articles/009763/2005-01-26/>

St Louis University (1912), *The Geophysical Observatory.* Bulletin of St. Louis Observatory, Meteorology in St Louis University, 8 (1) April 1912 pp. 55. <http://www.eas.slu.edu/eqc/eqc_history/Bulletin1911/SLU_Bull_vol8.1_p1_Apr1912.pdf)>

Tissandier, A. (1891), 'Souvenirs d'un voyage autour du monde. Shanghai et Zi-ka-weï'. *La Nature* 944: 75-78.

Udiás, A. (1996), 'Jesuits' contribution to Meteorology'. *Bulletin of the American Meteorological Society* 77: 2307–2315.

Udiás, A. (2003), *Searching the heavens and the Earth: the history of the Jesuit observatories*. Astrophysics and Space Science Library, 286. Kluwer Academic Publishers, Dordrecht. pp.369.

Udiás, A. and W. Stauder (1996), 'The Jesuit contribution to seismology'. *Seismological Research Letters* 67 (3): 10-19. <https://doi.org/10.1785/gssrl.67.3.10>

Whitaker, R. J. (2001), 'Harmonographs II. Circular design'. *American Journal of Physics* 69 (2): 174-183. <https://doi.org/10.1119/1.1309522>

Williamson, F. and C. Wilkinson (2017), 'Asian extremes: Experience and exchange in the development of meteorological knowledge c.1840-1930'. *History of Meteorology,* 8: 159-178.
< https://journal.meteohistory.org/index.php/hom/issue/view/9>

WMO (2012), 'The 140th Anniversary of the Shanghai Xujiahui Observatory'. <https://public.wmo.int/en/media/news-from-members/140th-anniversary-of-shanghai-xujiahui-observatory>

WMO (2021). 'Centennial Observing Stations'. <https://public.wmo.int/en/our-mandate/what-we-do/observations/centennial-observing-stations>

Zhu, M. (2012), *Typhoons, meteorological intelligence, and the inter-port mercantile community in nineteenth-century China.* PhD dissertation, Binghamton University, State University of New York, NY. pp. 322.

Zurcher, F. (1884), 'Les typhons. Observations de M. Le P. Dechevrens'. *La Nature* 594: 306-308.

Appendix

Works by Père Marc Dechevrens

A chronological list of known works by Marc Dechevrens compiled from diverse sources and not necessarily complete. It has not been possible to verify the full reference for some items.

1874 Observations faites à Shang-Haï. *Société Météorologique de France*, 7 (1) pp.72-74.

1876 Observations magnétiques faites à Zi-ka-weï. *Société Météorologique de France*, 9 (1) pp.67-70.

1876 Note sur l'installation de l'Observatoire magnétique de Zi-ka-weï'. *Société Météorologique de France*, 9 (1) pp.71-72.

1876 Osservazioni magnetiche fatte nell' Osservatorio di Zi-ka-weï. Roma, *Bull. Meterol.*, 15, pp.25-26.

1876 Recherche de l'influence de la lune sur les mouvements de l'aiguille de déclinaison à Zi-ka-weï. Roma, *Bull. Meteorol.* 15, pp.26-27.

1876 Recherches sur les principaux phénomènes de météorologie et de physique terrestre a l'Observatoire météorologiques et magnétique de Zi-ka-weï, près de Chang-haï (Chine). L'Annuaire de la *Société Météorologique de France*, 24, pp.189-278, 55 Tables, 11 plates. <https://books.google.je/books?id=0Xk9AQAAMAAJ&pg=RA2-PA13&source=gbs_selected_pages&cad=2#v=onepage&q&f=false>

1876 Recherches sur les principaux phénomènes de météorologie et de physique terrestre à l'Observatoire de Zi-ka-weï. (Extrait de l'Annuaire de la *Société Météorologique de France*, 24, pp.92, 11 plates (Separate issue of the above paper? Note title change).

1877 Recherches sur la variation des réguliers vents à Zi-ka-weï, Chine, après les observations faites de 1873 à 1877. Zi-ka-weï, pp.25, 8 plates (Available from NOAA Library, USA).

1877 La lumière zodiacale en chine, 5 années d'observations. L'Annuaire de la *Société Météorologique de France*, 24, pp.28, 6 tables.

1878 Calendier pour l'an 1878. Zi-ka-weï, pp.55, 8 plates.

1879 La lumière zodiacale étudiée d'après les observations faites de 1875 à 1879 à l'Observatoire de Zi-ka-weï, Zi-ka-weï, pp.II+38, 2 plates.

1879 Lettre relative aux observations faites l'Observatoire de Zi-ka-weï (Chine). L'Annuaire de la *Société Météorologique de France*, 27, 128-131.

1879 Le Typhon de 31 Juillet 1879. Zi-ka-weï, pp.iv+33+XVII. (Available from NOAA Library, USA).

1879 The typhoon of July 1879. Zi-ka-weï, pp.8+XV, 4 plates (English summary of the previous paper).

1879 Typhons des mois d'Août, Septembre, Octobre et Novembre 1879. (Appendix to the above).

1880 On the storms of the Chinese seas and on the storm of the 19th and 20th March 1880. Zi-ka-weï, pp.16, 3 plates.

1880 Instructions in the use of meteorological instruments for observers in China. Zi-ka-weï, pp.II+65, 8 plates.

1880 La Lumiére Zodiacle. Extracts of a memoir published under 'Notes and Extracts' in *Monthly Weather Review*, 8 (12), pp.17-19.

1880 Magnetic storm at Zi-ka-weï, near Shanghai, China, in August 1880. Extracts of a memoir published under 'Notes and Extracts' in *Monthly Weather Review*, 8 (12), pp.17-19.

1880 La pertubazione magnetica staordinaria dell' 11-14 agosto 1880. *Pontif. Univ. Gregor.* 19, p.84, pp.89-92.

1881 The typhoons of the Chinese Seas in 1880. Zi-ka-weï, pp.34, 2 plates.

1881 Correspondance sur la théorie des cyclones. *La Nature* 614, p.30.

1881 Sur l'inclinaison des vents. Nouvelle girouette pour observer cette inclinaison. Zi-ka-weï, pp.I+32, 9 plates (Available from NOAA Library, Washington).

1881 Le magnétisme terrestre à Zi-ka-weï, Chine. Zi-ka-weï, pp.II+171, 5 plates.

1881 The climate of Shanghai. Its meteorological condition. pp.231-248. (Unknown journal).
<http://docs.lib.noaa.gov/rescue/ cd128_pdf/LSN0764.PDF>

1882 The typhoons of the Chinese Seas in the year 1881. Zi-ka-weï, pp.II+171, 5 plates.

1882 Meteorological service on the coast of China. Instructions for keeping the meteorological log. Zi-ka-weï, pp.III+13.

1882 A curious halo. *Nature*, 27, (9 November) pp.30-31.

1882 Observation magnétique faite pendant l'éclipse totale de Soleil du 17 mai. *Astronomie*, pp.270-271.

1883 Régime des vents à Zi-ka-weï de 1877 à 1882. Zi-ka-weï.

1883 Variations de l'aiguille aimantée pendant les éclipses de lune – Régime des vents à Zi-ka-weï. 1877-1882. Zi-ka-weï, pp.31, 3 plates.

1884 Les typhons de 1882 – 1ere partie: les typhons des mois Juillet (5) et Août (5). Zi-ka-weï, pp.54, 7 plates.

1884 The typhoons of 1882 – Part II: typhoons in September (2) and October (2). Appendix – Storm of 16th and 17th May 1882. Zi-ka-weï., pp.32, 2 plates.

1885 The meteorological elements of the climate of Shanghai. Twelve years of observations made at Zi-ka-weï. Zi-ka-weï, pp.35.

1885 Bolide extraordinaire observé en Chine. (Correspondance). *La Nature* 614, p.214.

1885 L'humidité de l'air. Cause d'erreur du thermomètre à boule mouillée. *La Nature* 615, p.251.

1885 Mouvements des couches élévees de l'atmosphère à Zi-ka-weï déterminés par la direction de Cirri. Zi-ka-weï, pp.15.

1885 The typhoons of the Chinese Seas in the year 1885. Zi-ka-weï, pp.42, 18 plates (Combined with following item).

1885 Essay on the atmospheric pressure variations in the Far East during January 1885 (62 small charts). Zi-ka-weï, (Combined with above item).

1885 L'inclinaison des vents sur l'horizon; un anémomètre (moulinet à palettes inclinées) pour observer cette inclinaison. Avec un appendice sur les courants verticaux dans les cyclones. 2ème note. Zi-ka-weï, pp.48, 3 plates.

1885 Sur une trombe observée à Shanghaï, le 21 août 1885. *Comptes Rendu – Académie des Sciences*, 101, pp.759-760.

1885 Ueber die Regenverhältnisse in China im Juni 1885. Wien. *Meteorologische Zeitschrift* 20, pp.414-7.

1886 L'inclinaison des vents sur l'horizon. 3eme note Première année d'observation, 1886. Zi-ka-weï, pp.35, 7 plates.

1886 Sur les courants verticaux dans les cyclones; constitution des cyclones et quelques conséquences qui en découlent. Zi-ka-weï.

1886 On vertical currents in cyclones. *American Meteorological Journal*, 3 (4), pp.170-84. (Translation of 'Sur les courants . . . see above)

1886 Typhoon in the China Sea Aug 14, 1886. *Nature*, 34, p.578.

1886 La pluie d'étoiles filantes du 27 novembre 1885 à l'Observatoire de Zi-ka-weï, près de Changhai (Chine). *Comptes Rendus - Académie des Sciences*, 102, p.307.

1887 Sur la variation de température observées dans les cyclones. 2eme note. Zi-ka-weï, pp. 17.

1887 L'inclinaison des vents sur l'horizon de Zi-ka-weï pendant l'année 1886. Zi-ka-weï.

1887 Sur la reproduction expérimentale des trombes. *Comptes Rendus - Académie des Sciences*, 105, pp.1286-1289.

1887 Réponse à M. Faye sur la critique qu'il a faite [des expériences de l'auteur] sur les trombes artificielles. *Comptes Rendus - Académie des Sciences*, 106, pp.222-225.

1888 Quel est le sens des courants verticaux au centre des cyclones? *Comptes Rendus - Académie des Sciences*, 106, pp.1303-1306.

1888 Variation diurne de l'inclinaison des mouvements de l'air observées à Zi-ka-weï en Chine. *Comptes Rendus - Académie des Sciences*, 106, pp.1697-1700.

1888 Notice sur l'Observatoire de Zi-ka-weï. *Études*, Paris, 25, pp.18.

1888 Influence de l'altitude sur les variations de la température dans leur relation avec la pression. *Cosmos*, Paris, p.3.

1888 La température dans les cyclones d'après les observations de plaines et de montagnes. *Cosmos*, Paris.

1888 Les tourbillons atmosphériques, leur formation, leur constitution. *Memorie della Pontifica Academia dei Nouvi Lincei*, Rome, 4, pp.19-40.

1889 Influence de l'altitude sur les variations de la température dans leur relation avec la pression. *Cosmos*, Paris, (mars) p.4.

1889 Comparaisons des températures d'été en plusieurs stations de montagnes françaises avec celles de Paris. *Cosmos*, Paris.

1889 Notice sur l'Observatoire de Zi-ka-weï, Chine. *Études*, Paris.

1890 The anemometer for vertical components. *American Meteorological Journal*, 6, pp.578-579.

1890 La méthode de calcul trigonométrique de Bessel transformée en une méthode de calcul arithmétique. *Ass. Franç, Comptes Rendu*, (Pt. 2), pp.300-307.

1890 Nouvelle méthode de calcul pour interpolation et la correction des observations météorologiques. *Comptes Rendus - Académie des Sciences*, 110, pp.1021-1024.

1890 Sur la variation de la température avec l'altitude dans les cyclones et les anticyclones'. *Comptes Rendus - Académie des Sciences*, 110, pp.1255-1258.

1890 Compte-rendu d'un ouvrage de M. J. de Sugny, lieutenant de vaisseau: 'Eléments de météorologie nautique'. *Études*, Paris, (1890) XXVII année.

1892 Méthode chronométrique pour calculer l'altitude et la vitesse de déplacement des nuages suivie d'une méthode simple pour déterminer l'inclinaison de leur trajectoire sur l'horizon. *Memorie della Pontifica Academia dei Nouvi Lincei*, Rome, 8, pp.73-84.

1893 Les tremblements de terre, à l'occasion d'un livre de Léon Vinot sur la question. *Études*, Paris, 30, p.9.

1894 La bourrasque des 10 et 11 juillet 1894. *Cosmos*, Paris.

1894 La bourrasque des 10 et 11 juillet 1894 dans la Manche. *Comptes Rendus - Académie des Sciences*.

1894 Le campylographe. *Société Française de Physique*, pp.223-8 (3 figures).

1895 Le clino-anémomètre. *Société Météorologique de France*, 43, pp.56-61.

1895 Pression et température dans les cyclones d'Europe. *Société Météorologique de France*, 43, pp.219-20.

1895 Mouvements des diverses couches de l'atmosphère'. *Memorie della Pontifica Academia dei Nouvi Lincei*, Rome, 11, pp.245-251.

1895 Sur la composante verticale du vent. *Société Météorologique de France*, (Décembre).

1896 Mouvements des diverses couches de l'atmosphère. *Memorie della Pontifica Academia dei Nouvi Lincei*, Rome, 11, pp.11, 1 plate.

1896 Sur les hautes pressions atmosphériques de Janvier 1896. *Comptes Rendus - Académie des Sciences*, 122, pp.82-4.

1897 Clima di Jersey (Isola sul Canale della Manica). *Monccalieri Oss. Boll.*, 17, p.90.

1897 Le mouvement oscillatoire diurne de l'atmosphère. *Comptes Rendus - Académie des Sciences*, 124, pp.1479-1480.

1898 Das neue Observatarorium auf Jersey. *Meteorologische Zeitschrift*, 15, pp.149-50.

1898 Les variations de la température de l'air dans les tourbillons atmosphériques et leur véritable cause. *Revue des Questions scientifiques*, Bruxelles, 44, pp.521-552.

1898 Les variations de la température de l'air dans les cyclones, et leur cause principale. *Memorie della Pontifica Academia dei Nouvi Lincei,* Rome, 14, p.233-267.

1898 Les nuages dits en côtes de melons. *Cosmos*, Paris, (août).

1899 Les variations de la température de l'air dans les tourbillons atmosphériques et leur véritable cause. *Revue des Questions scientifiques*, Bruxelles, 45, pp.409-434.

1899 Méthode simplifiée dite des facteurs pour le calcul des séries de Fourier et de Bessel appliquées à la météorologie. *Memoire della Pontifica Academia dei Nouvi Lincei*, Rome, 16, pp.149-182.

1900 Note complémentaire à la méthode simplifiée dite des facteurs pour le calcul des séries de Fourier et de Bessel appliques spécialement à la météorologie. *Memoire della Pontifica Academia dei Nouvi Lincei*, Rome, 17, pp.47-64 (2 plates).

1900 Le camplylographe, machine à tracer des courbes. *Comptes Rendus - Académie des Sciences*, 130 (11 Juin) pp.1616-1620.

1900 Vision stéréoscopique des courbes tracées par les appareils phases. *Comptes Rendus - Académie des Sciences*, 131, (15 Août) pp.408–410.

1900 Deux notes sur la campylographe et sur les figures stéréoscopiques. *Comptes Rendus - Académie des Sciences*.

1900 La variation de la température dans les cyclones selon l'altitude, d'après les observations de montagnes. Paper read at the *Congrès International de Météorologie*, Paris.

1900 Collection de 25 figures stéréoscopiques. Editeur, la Maison Dubosq - M. Pellin de Paris.

1901 Le campylographe: Appareil à dessiner des courbes géométriques, des figures stéréoscopique et des dessines artistiques. *Revue des Questions Scientifiques*, Bruxelles, 2ème séries, 19, pp. 21-46, with plates.

1901 Le Campylographe. Polleunis & Ceuterick, Louvain, pp. 27.

1901 Sur les causes des variations accidentelles de la température de l'air. *Société Météorologique de France*, (mai).

1902 Sur la génération de quelques courbes remarquables par la campylographe (en collaboration avec le P. Potron). *Revue des Questions scientifiques*, Bruxelles, p.17.

1902 La campylographe Dechevrens et le compas Schwartzbard. *Cosmos*, Paris, (Février).

1902 Les variations passagères de la température, causes ou effets des tourbillons atmosphériques? *Memoire della Pontifica Academia dei Nouvi Lincei*, Rome, 19, p.32.

1902 Variations passagères de la température, cause ou effet des tourbillons. *Symon's Meteorological Magazine*, London (A review of the previous paper).

1902 Le température à Jersey, le maximum d'été. *Société Astronomique de France.*

1903 Réchauffements et refroidissements locaux à la date du 16 avril 1903. *Société Météorologique de France*, (mai).

1904 The vertical component of the wind. *Monthly Weather Review* 32 (3), pp.118-210.

1904 Sur quelques variations intéressantes de la température en Europe, les 3 et 4 décembre 1903. *Société Météorologique de France*, (janvier).

1904 Sur le refroidissement survenu en France le 23 janvier 1904. *Société Météorologique de France*, (février).

1904 Les variations de la température selon la verticale: une seconde altitude limite. *Société Météorologique de France*, (mars).

1904 The Dechevrens anemometer. Cold waves. Translation of extracts of a Dechevrens letter. *Monthly Weather Review* 32 (10), p.472.

1904 La théorie hydro-thermodynamique (ou mécanique) des tourbillons atmosphériques. Congrès de *l'Association Française pour l'Avancement des Sciences*, Grenoble.

1905 *La théorie hydrothermodynamique des tourbillons atmosphériques.* Jersey. pp.35 (Written in response to reaction to the previous paper).

1905 La théorie hydrothermodynamique des tourbillons atmosphériques. *Symon's Meteorological Magazine*, London. (Review of the paper above).

1905 La théorie hydrothermodynamique des tourbillions. *Bolletino della Societa Meteorologica Italiana*. (Italian translation of the above).

1905 La théorie hydrothermodynamique des tourbillons. (Spanish translation by M. Leal of Léon Observatory, Mexico).

1905 La variation diurne de la tension de la vapeur d'eau à Jersey. *Société Météorologique de France* (avril).

1906 La pression et la température de l'air dans les cyclones et les anticyclones. Réponse à M. J. Hann. *Bulletin des Observations Magnétiques et Météorologiques. XXIIème année 1905.* Jersey, pp.1-7.

1906 L'inclinaison du vent sur l'horizon de Jersey. *Memoire della Pontifica Academia dei Nouvi Lincei*, Rome, 24, pp.3 (3 plates).

1906 L'éruption de Vésuve observée à Jersey. *Revue Nephoscopique*, Mons.

1906 A propos des causes du refroidissement de l'air. *Société Belge d'Astronomie*, Bruxelles.

1906 Sur la cause des refroidissements accidentels. *Société Belge d'Astronomie*, Bruxelles.

1906 La pression et la température de l'air dans les cyclones et les anticyclones, réponse à M. Hann. *Bolletino della Societa Meteorologica Italiana*. (Italian translation of previous paper).

1906 La radiation terrestre par ciel découverte, est-elle la principale cause de refroidissement de l'air? *Société Belge d'Astronomie*, Bruxelles, 5.

1907 La variation diurne de la tension de la vapeur d'eau à Jersey de 1894 à 1903, et en 1906. *Société Météorologique de France*, (mars).

1907 Observation de Mercure à l'oeil nu (Note). *Société Astronomique de France* 21, p.60.

1907 Le mouvement de Vénus par rapport à la terre (tracé par le Campylographe-Dechevrens et vu dans l'espace à l'aide du Stéréoscope). *Bulletin de la Société Astronomique de France*, 21, pp.96-98 (2 figures).

1907/8 Album de 300 dessins campylographiques. Héliogravure de Tillet, Paris: Stéréotypie de Nancy, (1907 et 1908).

1908 Observations actinométriques à Jersey. *Société Météorologique de France*, (mars).

1908 Froide et brouillard sur la Manche du 16 au 20 Mai 1908. *Société Météorologique de France*, (juin).

1908 Les phénomènes de température dans les tourbillons et en particulier dans la haute atmosphère. *Memoire della Pontifica Academia dei Nouvi Lincei*, Rome, 26, pp.34.

1908 Les phénomènes de température dans les tourbillons et en particulier dans la haute atmosphère. *Bolletino della Societa Meteorologica Italiana*. Turin. (Translation of the previous paper).

1909 Les phénomènes de température dans les tourbillons et en particulier dans la haute atmosphère. *Revue Nephoscopique*, Mons. (Copy of the previous paper).

1909 Sur le chant des lignes télégraphiques. *Société Météorologique de France*, (Janvier).

1909 La fréquence des halos à Paris et à Jersey. *Revue Nephoscopique*, Mons.

1909 Un bolide remarquable. *Cosmos*, Paris, (février).

1911 Sur un phénomène d'inversion de la température dans la Mer Egée. *Société Météorologique de France*, (Novembre).

1911/1912 La méthode de prévision du temps de M. G. Guilbert expliquée par la théorie hydro-dynamic des tourbillons. *Memoire della Pontifica Academia dei Nouvi Lincei*, Rome, 65, p.9.

1913 Les ondes hertziennes atmosphériques enrégistrées à Jersey de décembre 1911 à avril 1913. *Memoire della Pontifica Academia dei Nouvi Lincei*, Rome, 24.

1913 Warm and cold winds. *Symon's Meteorological Magazine*, London.

1914 Les ondes hertziennes atmosphériques enrigistrées et étudiées à l'observatoire Saint-Louis, Jersey en 1912 et 1913. *L'Association Française pour l'Avancement des Sciences*, Congrès de Havre, (1914) pp.314-23.

1914 Les ondes hertziennes atmosphériques observées à Jersey en 1912 et 1913. *Revue des Questions Scientifiques*, Bruxelles.

1916 Les ondes hertziennes atmosphériques. *Terrestrial Magnetism and Atmospheric Electricity*, Washington, (Juin).

1917 Les courants telluriques étudies par la méthode simple à l'Observatoire Saint-Louis à Jersey. *Memoire della Pontifica Academia dei Nouvi Lincei*, Rome, 3, pp.36.

1918 Contribution à l'étude de la variation diurne de l'électricité atmosphérique. *Memoire della Pontifica Academia dei Nouvi Lincei*, Rome, 4, pp.47.

1918 Results of observations of earth currents made at Jersey, England, 1916-1917. *Terrestrial Magnetism and Atmospheric Electricity*, Washington, 23 (1), pp.37-39.

1918 Additional results of earth-current observations at Jersey, England. *Terrestrial Magnetism and Atmospheric Electricity*, Washington, 23 (3), pp.145-147.

1918 Le courant tellurique comme marée électrique. *Comptes Rendus - Académie des Sciences*, (Octobre).

1919 Similarities between the earth current observed at Jersey, England, and an electric tide as derived from the ocean tide. *Terrestrial Magnetism and Atmospheric Electricity*, Washington, 24 (1), pp.33-38.

1919 Les variations diurnes du courant électrique vertical de la terre à l'air. *Comptes Rendus - Académie des Sciences*, 168 (mars).

1919 Sur les points du globe où il se produite une marée électrique, dérivée de la marée océanique, observe-t-on une marée magnétique? (Letter to the Editor). *Terrestrial Magnetism and Atmospheric Electricity*, Washington, 24 (4), pp.175-177.

1919 Concerning the electric tide observed at Jersey'. (Letter to the Editor). *Terrestrial Magnetism and Atmospheric Electricity*, Washington, 24 (4), pp.178-179.

1920 Modification et complément apportées à la méthode d'observations tellu-riques à l'aide de conducteurs nus souterraines. *Comptes Rendus - Acadé-mie des Sciences*, 169.

1920 Carrière scientifique du P. Marc Dechevrens, de la Compagnie de Jésus. (Typescript, Maison St. Louis Observatory, Jersey with corrections in the hand of Dechevrens), pp.6.

c. 1920 Catalogue des publications diverses faites par le P. Marc Dechevrens S. J. en Chine et en Europe à partir de 1874. (MS in Dechevrens' hand, Maison St Louis Observatory, Jersey).

1921 Singulier phénomène de résonance électrique dans un circuit aérien. *Ciel et Terre*, Bruxelles, 37, pp.89-93.

1922 Le courant tellurique et la marée électrique. *Annales de la Société Scienti-fique de Bruxelles.*

1922 Singulier phénomène de résonance électrique dans la circuit aérien. *Société Belge d'Astronomie.*

1922 Deux catégories de courant tellurique. *Comptes Rendus - Académie des Sciences*, 175, (décembre).

1923 Deux catégories de courants telluriques. *Revue des Questions Scientifiques,* pp.302-25.

1923 Étude du vent à Jersey: vingt années d'observations (1895-1914) à l'Ob-servatoire Saint-Louis. *Mémorial de l'Office National Météorologique de France*, Paris, 1 (3), pp.55.

Undated (c. 1912) Contribution du Père Marc Dechevrens S. J. à l'étude de l'atmosphère principalement de ses mouvements tourbillonnaires et de ses température avec indication des travaux étrangers plus en relation avec ceux de l'auteur. MS, St Louis archive pp.14.

Appendix notes

Cosmos was a weekly science review. The full title of 'Études' is: Études religieuses, philosophiques, historique et littéraires des P.P. de la Compagnie de Jésus.

Regular reports produced by Dechevrens

1874 to 1886: Bulletins des observations magnétiques et météorologiques. Volumes 1 (1874 and 1875) to Volume 12 (1886). 12 reports of 175 to 240 pages and 2 to 24 plates in each volume.

1894 to 1913: Bulletins météorologiques et magnétiques des observations faites à l'observatoire à Jersey. 19 reports of 28 to 36 pages.

1914-1916: Bulletins météorologiques et magnétiques des observations faites à l'observatoire à Jersey. 3 reports of 40 pages published by the Meteorological Office, London.

1914-1920: Monthly summary published in the Annual Bulletin of the Société Jersiaise. Seven single page reports.

Periodic works

From 1895: Monthly résumés of the observations published in the monthly bulletin of the Bureau Central Météorologique de France.

From 1901: *Observations spéciales des nuages et du vent* made on 3 consecutive days each month published in the monthly report of the 'Commission Internationale d'Aerostation Scientifique' Strasbourg.

1894 to 1914: Weekly weather summaries for the Jersey Times and La Chronique newspapers.

1906 to 1907: Weekly weather summaries for the publication 'Jersey Week by Week'.

1894 to 1920: Annual Bulletin of the Société Jersiaise. 26 annual summaries of the St Louis observations.

1896: A short article on the climate of Jersey in the illustrated album 'Sunny Jersey'.

1899, 1902, 1906 and 1907: Short articles on the climate of Jersey in the publication 'Beautiful Jersey'.

1908: A short article on the climate of Jersey in the illustrated album 'Jersey, the Riviera of Great Britain'. (French and English editions).

Other works and instruments

The St. Louis 50-metre metal tower – 1894. Architect M. Seyrig, Paris.

La Maison St. Louis Observatory – 1894. Builder Samuel Cuzner, 22 Great Union Road, Jersey.

L'Anémomètre-Universal Dechevrens for the tower – 1894. Built by *Richard Frères* of Paris.

Weathervane for the Observatory roof to record wind direction.

Large scale 'aéro-hydraulique' barometer.

An anemometer to record absolute wind speed, built by M. Chateau in Paris.

The first simple campylographe. Built by M. Chateau.

The second full scale campylographe. Built by M. Cretin-Lange of Moren, Jura.

An electro-magnetic wind vane for the tower to record at distance.

Sources used to compile Dechevrens list of works:

1. Lettres de Jersey 'Souvenir du cinquantaine 1880-1931'. Editor E. D. Livres et articles écrits à Jersey 1880-1930.

2. Anon (1901). The International Meteorological Congress, Paris, September 10-16, 1900. *Monthly Weather Review* 29 (6), pp. 265-8

3. Indices to the magazine 'La Nature' are available on web site: http://cnum.cnam.fr/CGI/redira.cgi?4KY28

4. Catalogue des publications diverses faites par le P. Marc Dechevrens S. J. (Dechevrens c. 1920)

5. Copies of papers held by the JMD.

6. Udiás, A. (2003) – see references.

7. Several random chance encounters.

Citations
Extensive quotation and use of Dechevrens work in:

J. de Sugny (1890) *Éléments de Météorologie Nautique*, Berger-Levrault et Cie, Paris et Nancy.

www.ingramcontent.com/pod-product-compliance
Lightning Source LLC
LaVergne TN
LVHW070059080426
835511LV00024B/3472